ESSENTIALS OF OCEANOGRAPHY

ESSENTIALS OF OCEANOGRAPHY

Jakub Trojan
Eva Janska

Kruger Brentt
Publishers

2025

Kruger Brentt Publishers UK. LTD.
Company Number 9728962

Regd. Office: 68 St Margarets Road, Edgware, Middlesex HA8 9UU

ISBN: 978-1-78715-363-9

For information on all our publications visit our website at http://krugerbrentt.com/

PREFACE

"Essentials of Oceanography" embarks on a fascinating exploration of the dynamic world of the oceans, offering readers a comprehensive understanding of the fundamental processes, features, and interactions that shape Earth's marine environments. This book serves as an indispensable resource for students, educators, researchers, and enthusiasts seeking to grasp the essentials of oceanography and its significance in understanding our planet's oceans.

The oceans cover over 70% of Earth's surface and play a vital role in regulating the planet's climate, supporting biodiversity, and sustaining life. From the deep abyssal plains to the vibrant coral reefs, the oceans harbor a diverse array of ecosystems and habitats that are interconnected through complex physical, chemical, and biological processes.

The primary objective of this book is to provide readers with a comprehensive overview of oceanography, covering key topics such as ocean basins and features, ocean circulation, waves and tides, marine ecosystems, marine biodiversity, ocean chemistry, and the role of the oceans in climate regulation. Each chapter offers insights, scientific principles, and real-world examples to help readers understand the dynamic nature of Earth's oceans and the interdisciplinary nature of oceanographic research.

Contributions from leading experts in the field explore various aspects of oceanography, including oceanographic exploration techniques, satellite remote sensing, marine conservation efforts, and the impacts of human activities on marine ecosystems. Special emphasis is placed on the importance of interdisciplinary collaboration and global cooperation in addressing pressing ocean-related challenges such as climate change, pollution, overfishing, and habitat destruction.

As our understanding of the oceans continues to evolve and the need for sustainable ocean management becomes increasingly urgent, it is essential for individuals and societies to appreciate the significance of oceanography in informing sound decision-making and stewardship of marine resources. By fostering ocean literacy and promoting scientific inquiry, we can work together to protect and preserve the invaluable ecosystems and resources of our oceans for future generations.

We extend our sincere gratitude to the contributors whose expertise and dedication have enriched this book. It is our hope that "Essentials of Oceanography" will serve as a valuable resource and inspiration for readers interested in exploring the wonders of the oceans and understanding their profound impact on the health and well-being of our planet.

Jakub Trojan
Eva Janska

CONTENTS

Chapter 1 INTRODUCTION

1.1 OCEANS: AN OVERVIEW

To start, let's examine some fundamental information regarding the oceans. When considering Earth, we commonly focus on its land extent. However, it is important to note that oceans actually encompass 71% of the Earth's surface, while land accounts for only 29%. The Earth's oceans span an expanse of around 139 million square miles or 361 million square kilometers, and have a volume of over 1.37 billion cubic kilometers of water. The distribution of water on Earth is not uniform. Oceans cover 61% of the surface area in the Northern Hemisphere, whereas in the Southern Hemisphere, they comprise 81% of the surface area (Figure 1.1.1).

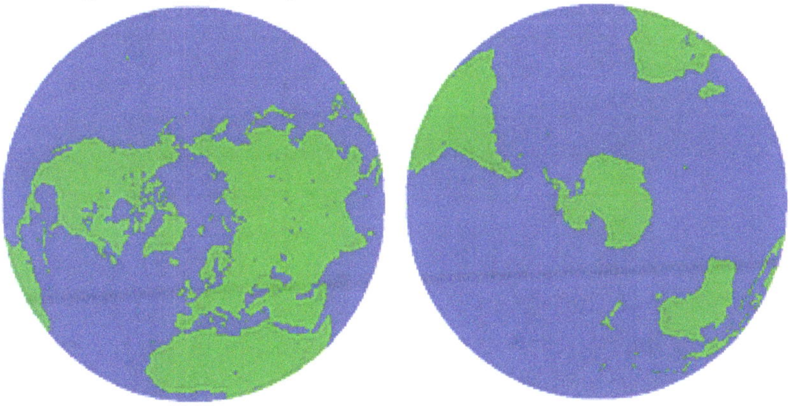

Figure 1.1.1 Ocean cover in the Northern (left) and Southern (right) Hemispheres.

Different sources vary in the number of acknowledged ocean basins. Traditionally, the primary oceans that have been acknowledged are the Pacific, Atlantic, Indian, and Arctic Oceans. In recent times, the Southern Ocean has been officially designated as the fifth named ocean. It includes all the water from the coast of Antarctica down to 60 degrees south latitude (Figure 1.1.2). In 2000, these limits were officially presented to the International Hydrographic Organization for approval. However, some countries do not acknowledge it as an independent ocean, but rather consider it as the southern extension of the other primary oceans. Due to its distinct characteristics, the Southern Ocean will be considered as a separate ocean in this book.

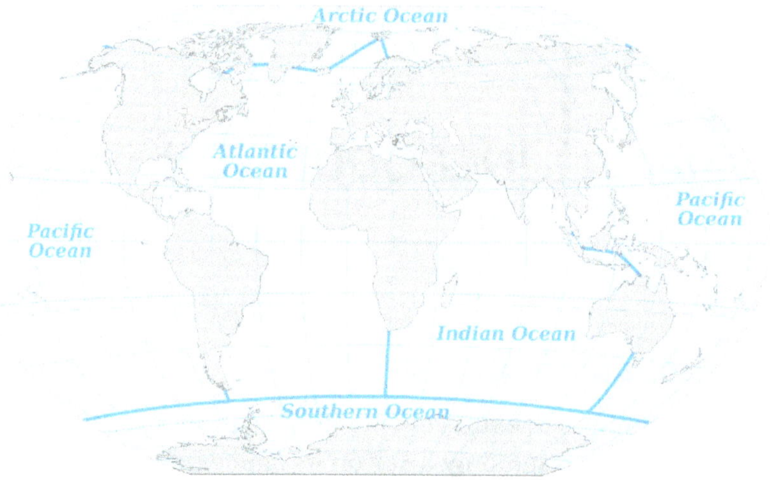

Figure 1.1.2 Map of the world oceans (By Pinpin [GFDL (http://www.gnu.org/copyleft/fdl.html) or CC BY-SA 3.0], via Wikimedia Commons).

The oceans comprise a significant portion of the Earth's water, encompassing 97% of the water found on the planet's surface. The Pacific Ocean alone has more than half of this water (as seen in Table 1.1.1).

Table 1.1.1 Percentage of Earth's water in various locations

Pacific	52%
Atlantic	25%
Indian	20%
Ice	2%
Ground water	0.6%
Atmosphere, lakes & rivers	0.01%

The mean depth of the global ocean is around 3800m (12,500 ft), making it roughly four times deeper than the average elevation of land, which is 840m or 2800 ft. Mount Everest, the highest terrestrial peak, stands at an elevation of 8848 meters (29,028 feet), whereas the Challenger Deep in the Marianas Trench, the deepest point in the ocean, reaches a depth of roughly 10,920 meters (36,200 feet). Even if Mt. Everest were placed in the Marianas Trench, it would still be submerged under more than 2 km of water. Due to the vast amount of water compared to land on Earth, if the land elevation were leveled, the entire planet would still be submerged under around 2700 meters of water.

The Pacific Ocean is the largest among the major ocean basins, roughly equal in size to the combined size of all the other basins. Additionally, it is the deepest ocean basin according to Table 1.1.2.

Table 1.1.2 Area and depth of the major oceans

	Area (million km^2)	Average depth (m)
Pacific	166	4282
Atlantic	87	3926
Indian	73	3963
Arctic	14	1205
Southern	20	4000

1.2 CONTINENTAL MARGINS

*Portions modified from "Physical Geology" by Steven Earle**

Continental margins are the areas where the land transitions to the deep bottom, specifically between the continental and oceanic crust. Within an active continental margin, the interface between the continent and the ocean coincides with a tectonic plate boundary, resulting in significant geological activity in the vicinity of the border. The western coast of the United States exemplifies an active margin, where the shoreline aligns with the boundary between the Pacific and North America Plates. A passive continental margin is a type of boundary between land and sea that is not connected to a tectonic plate boundary. An exemplary instance is the eastern coast of the United States, where the plate boundary lies along the mid-Atlantic ridge, situated at a considerable distance from the coastline. Passive edges have lower levels of geological activity. Figure 1.2.1 depicts a conceptual passive margin in an idealized manner. When analyzing this diagram, and similar ones, it is important to observe that there is a substantial vertical exaggeration. The depth scale represents around 5000 m, while the horizontal scale spans around 300 km. This creates an illusion of greater steepness in the features, which is not reflective of their true degree of steepness. The lower part of Figure 1.2.1 illustrates the appearance of a passive edge without the depicted exaggeration. In reality, the transition to depth is much more gradual.

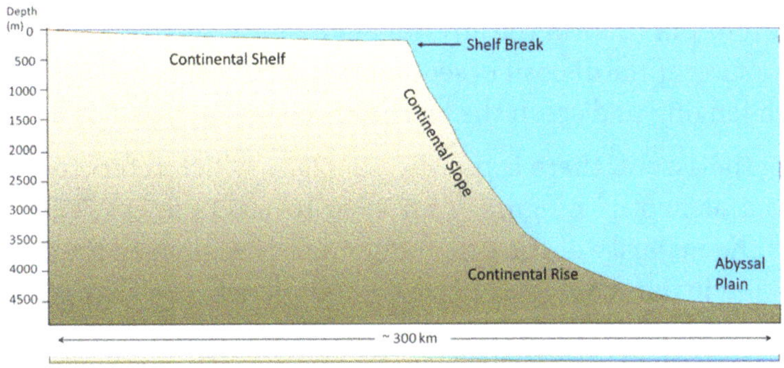

Figure 1.2.1 Features of a passive continental margin (modified by PW from Steven Earle, "Physical Geology").

The continental shelf refers to the relatively shallow and submerged area that forms the border of a continent. From a geological perspective, the shelf remains a constituent of the continental crust, although it is frequently covered by marine deposits. Typically, the distance from the shore to the shelf is approximately 80 km. However, certain areas have a narrow shelf, whereas the Siberian Shelf in the Arctic spans approximately 1500 km. The shelf's depth often does not exceed 150 meters, and its floor is relatively level. The flat terrain is a consequence of fluctuations in sea level. Over time, the shelves have alternated between being buried and exposed. As sea level has fluctuated, the forces of wave action, ice sheets, and other erosional processes have leveled the surface of the shelf. The process of smoothing has been ongoing due to wave action and the migration of sediments over the shelf. Although continental shelves comprise just approximately 6% of the total surface area of the ocean, they are biologically one of the most abundant regions. This is due to their relatively shallow depth, which prevents nutrients from sinking, and their close proximity to the coast, which results in a huge influx of nutrients. The continental shelf terminates at the shelf break, which designates the location where the inclination of the seafloor starts to increase more sharply. The average depth of the shelf breach is around 135 meters.

Following the shelf break, the seafloor exhibits an increased gradient (about 4 degrees) as it drops into the abyssal depths of the ocean. The continental slope, which is the steeper part of the margin, stretches from the shelf break to depths of 3000-5000m. Certain regions of the ocean exhibit the presence of expansive undersea canyons that have been eroded into the continental slope. One notable example is Monterey Canyon located in Monterey Bay, California, which has a comparable magnitude to the renowned Grand Canyon. Turbidity currents, which are essentially landslides of silt, rocks, and other debris, have the potential to carve out these canyons.

The continental rise is located at the base of the hill. This region denotes the boundary between the continental crust and the oceanic crust, where the incline gradually flattens out to form the abyssal plain of the deep ocean floor. The rise is composed of a substantial deposit of sediment that has collected from the continent, making it challenging to discern the boundary between the slope and the rise.

Following the ascent, there is the abyssal plain, which refers to the submerged seabed located at depths ranging from 4500 to 6000 meters. The abyssal plain encompasses the majority of the ocean floor and is renowned for its unparalleled flatness, making it the most level place on the planet. The flatness of the area is a result of silt accumulation over millions of years, which has covered up many structures on the bottom (Figure 1.2.2).

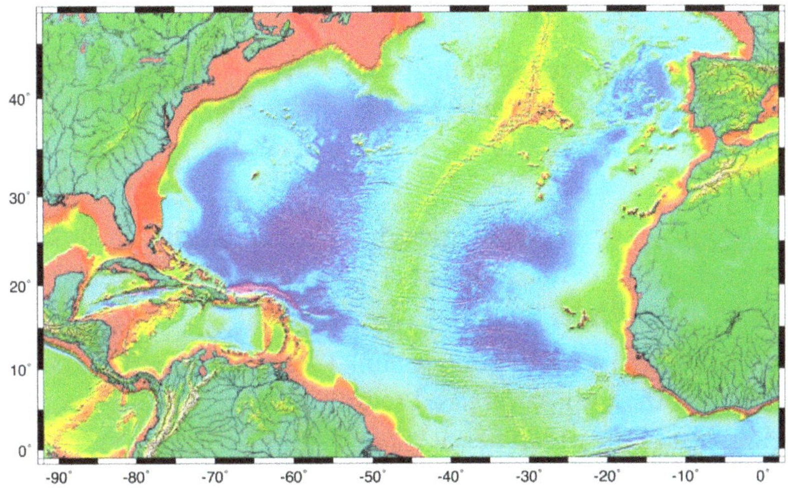

Figure 1.2.2 Topography of the North Atlantic. Shallow continental shelf regions are shown in red, and the abyssal plain is shown in blue. Along the east coast of the United States the continental slope can be seen in green (Steven Earle, "Physical Geology").

Passive margins, as previously explained, possess expansive shelves, gradual inclines, and a prominently formed rise. Passive margins, being distinct from plate borders, undergo extended periods of relative stability, which might result in the formation of these characteristics. Active margins exhibit comparable characteristics to passive margins, however, the parameters of these features are influenced by the plate boundary. Active edges, such as the Pacific coast of North America, are characterized by thinner shelves, steeper slopes, and minimal or no elevation, especially in areas where convergent borders occur. Trenches linked to subduction zones function as sedimentary reservoirs, impeding the formation of a continental rise and inhibiting the deposition of sediments onto the abyssal plains.

1.3 MARINE PROVINCES

In section 1.2, we acquired knowledge about the geographical areas that constitute the continental margins. Prior to concluding this discussion, we shall examine further methods for classifying ocean ecosystems.

The primary differentiation lies in the pelagic and benthic zones. The pelagic zone encompasses the entirety of the water column and serves as the habitat for species that swim or float. The benthic zone pertains to the lowest part of a body of water, while the organisms inhabiting and existing within this area are referred to as the benthos.

The pelagic zone is partitioned into two provinces: the neritic province encompasses the water from the low tide line to the shelf break, while the oceanic province encompasses all other water in the open ocean regions.

The oceanic province is divided into depth zones (Figure 1.3.1):

- 0-200 m is the **epipelagic** zone ("epi" = "upon", as in on top of the pelagic zone). This is the region where enough light penetrates the water to support photosynthesis (see section 7.3), so it is also called the **euphotic** or **photic** zone.
- 200-1000 m is the **mesopelagic** zone ("meso" = "middle"). There is some light here, but not enough for photosynthesis, so it is called the **dysphotic zone**, or the **twilight zone**.
- 1000-4000 m is the **bathypelagic** zone ("bathy" = "deep"). There is no light at these depths, so it is referred to as the **aphotic** zone. About 75% of the living space in the ocean lies at these depths or deeper.
- 4000-6000 m is the **abyssopelagic** or **abyssalpelagic** zone, which extends to the seafloor in most areas.
- 6000 m and below is the **hadopelagic** or **hadalpelagic** zone (named for Hades or "hell"). This refers to the water in deep ocean trenches.

Inhabitants of these regions are referred to according to their habitat, for example mesopelagic fish, or epipelagic squid.

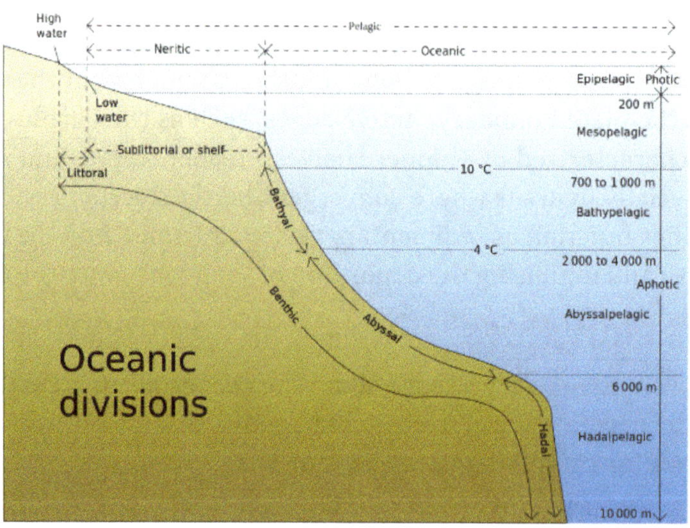

Figure 1.3.1 The major benthic and pelagic oceanic divisions (K. Aainsqatsi at en.wikipedia [Public domain, GFDL (http://www.gnu.org/copyleft/fdl.html) or CC-BY-SA-3.0], via Wikimedia Commons).

The benthic environment is also divided into zones, most of which correspond to the pelagic divisions:

- The **supralittoral** zone lies above the high tide line. Also called the spray zone, it is only submerged during storms or unusually high waves.

- The **littoral** zone is the region between the high and low tides. Thus it is also referred to as the **intertidal** zone.
- Below the littoral zone is the **sublittoral** (shelf) zone, extending from the low tide mark to the shelf break, essentially covering the continental shelf.
- The **bathyal** zone extends along the bottom from the shelf break to 4000m, so it generally includes the continental slope and rise.
- The **abyssal** zone is found between 4000-6000 m, including most of the abyssal plains. The abyssal zone represents about 80% of the benthic environment.
- The **hadal** zone includes all benthic regions deeper than 6000 m, such as in the bottom of trenches.

1.4 MAPPING THE SEAFLOOR

The preceding sections provided details regarding the oceanic depths in different locations. How can we accurately map the ocean floor to determine its immense depths?

Figure 1.4.1 Lead line survey from a catamaran hull in Alaska, 1942 (http://celebrating200years.noaa.gov/transformations/hydrography/image7.html).

In order to cartographically represent the ocean floor, it is important to ascertain the precise depth measurements at various locations. Bathymetry is the term used to describe the method of measuring the depths. The initial measurements were conducted using soundings, which involved lowering a weighted line (known as a lead line) by hand until it made contact with the seabed. The depth was then determined by measuring the length of the line (see Figure 1.4.1). This method resulted in the establishment of the fathom as a measurement unit for depth.

Sailors would extend the sounding line to the length of their arm span while pulling it in. The average arm span of a sailor was approximately six feet. Therefore, one fathom, which is equivalent to six feet, could be measured by counting the number of "arm spans" as the men pulled in the rope.

This method has some disadvantages and was typically restricted to more modest depths of water. The process was highly time-consuming and provided depth data for only one specific location, hence requiring numerous separate soundings to accurately map a given area. Additionally, there is a possibility of errors occurring. When the weight is dropped into deep water, it becomes challenging to accurately detect when it reaches the bottom. This is because the weight of the line itself can lead it to continue sinking, and the presence of currents can cause the line to deviate from a vertical position. As a result, there is a risk of overestimating the depth. Subsequently, winches and robust steel cables were employed to explore greater depths. However, this approach failed to completely address the underlying challenges of the sounding process and also imposed the limitation of excessive equipment weight.

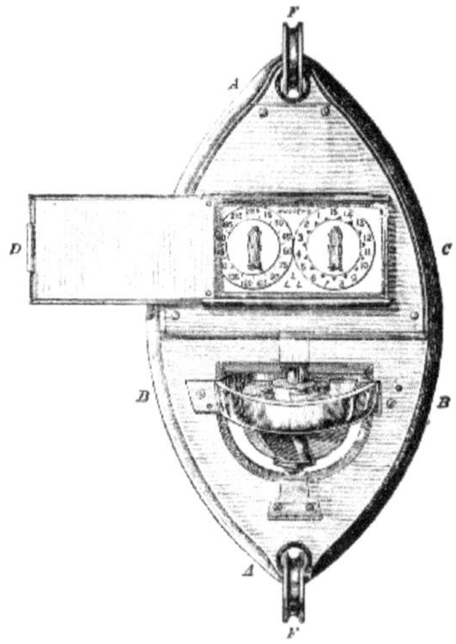

Figure 1.4.2 Massey's sounding machine (Public domain, via Wikimedia Commons).

Following the tragic event of the Titanic sinking in 1912, there was a concerted endeavor to enhance the techniques for identifying icebergs from a vessel. As a result, sonar technology, which stands for SOund Navigation And Ranging, was developed and quickly used for mapping bathymetry. An echosounder, which is a sonar equipment, emits a sound pulse and subsequently detects the echo that is reflected back. The time it takes for the echo to return is utilized to determine the depth. The speed of sound in water is roughly 1500 m/s (see to section 6.4). The water depth can be determined by multiplying half the time it takes for an echo to return by the speed of sound in water. This is because the returning echo travels to the bottom and returns (Figure 1.4.3).

During the 19th century, several alterations were implemented to enhance its uncomplicated form. In 1802, Edward Massey, a clockmaker from Britain, created a mechanical apparatus that was connected to the sounding line. As the apparatus descended, a rotor rotated a dial, which remained fixed when the line reached the bottom (Figure 1.4.2). Subsequently, the line could be retrieved and the depth could be determined by reading the measurement on the dial. In 1853, American seaman John Mercer Brooke invented a cannonball weight that was connected to a reel made of twine. The cannonball was released from the side and descended freely to the bottom. By measuring the pace at which the twine unwound and observing the change in rate upon impact with the bottom, the depth of the water could be determined. Upon reaching the bottom, the cannonball was dislodged and the line could be retrieved, along with a sample of mud adhered to the iron bar securing the projectile. This provided conclusive evidence that the bottom had indeed been reached.

depth=1/2*(two-way travel time)*(speed of sound in water)

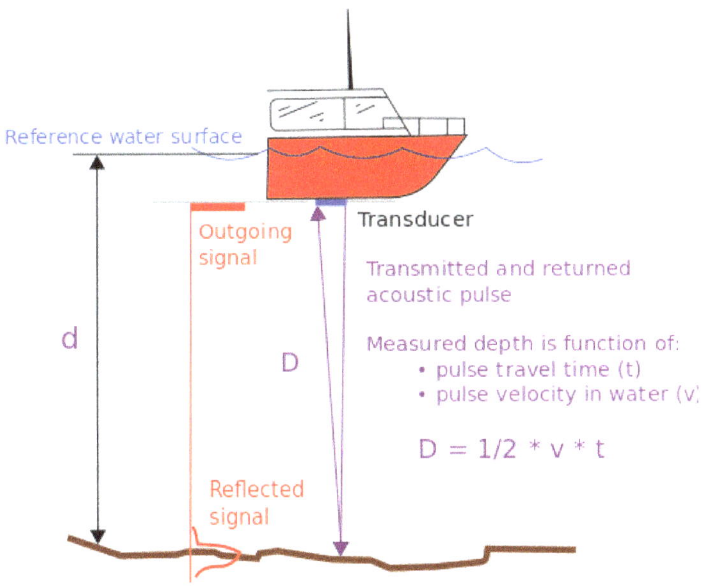

Figure 1.4.3 Measuring depth using an echosounder (Public domain via Wikimedia Commons).

Echosounders enabled the rapid and uninterrupted measurement of underwater depth profiles while a ship is in motion. However, they solely provide the depth exactly beneath the trajectory of a vessel. Currently, high-resolution seafloor maps are created using multibeam or side scan sonar technology, which can be operated either from a ship or a towed transmitter (as seen in Figure 1.4.4). Multibeam sonar generates a fan-shaped sonic field, enabling the simultaneous mapping of a significantly larger area (>10 km wide).

Figure 1.4.4 Multibeam sonar (NOAA).

Satellites, such as SEASAT, GEOSAT, and the Jason satellites, are used to conduct extensive mapping of the ocean floor. These satellites utilize radio waves to estimate the sea surface height through a technique called radar altimetry. The sea surface exhibits modest variations in elevation due to the influence of gravity, resulting in higher levels over elevated objects on the ocean floor and lower levels over trenches and other depressions.

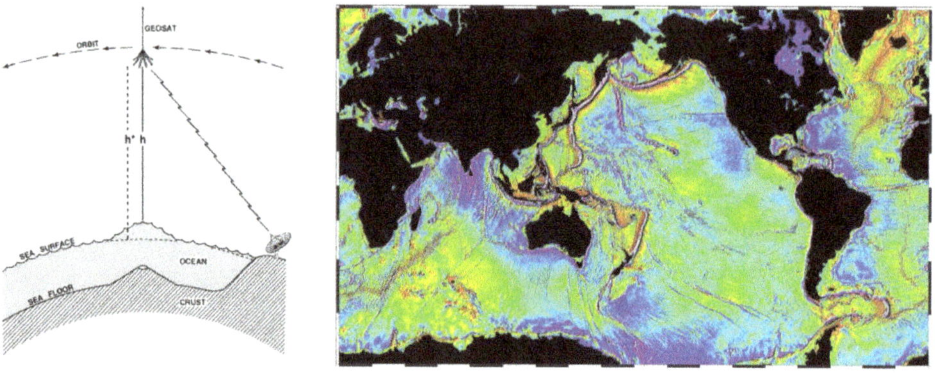

Figure 1.4.5 Radar altimetry (left) and a map of the seafloor produced by radar altimetry satellites (right) (NOAA).

Satellites emit radio waves and, like an echosounder, may utilize the reflected waves to measure variances in sea surface elevation with a precision of 3-6 cm (Figure 1.4.5). The variations in sea surface elevations enable us to ascertain the topography beneath the water's surface. Contrary to the outdated lead line method, which required numerous soundings to map a small area, with each sounding taking an hour or more to complete, the present satellites are capable of mapping more than 90% of Earth's ice-free sea surface every 10 days!

2 Chapter GETTING OUR BEARINGS

INTRODUCTION

For millennia, humans have been traversing the vast oceans, driven by the desire to explore, embark on journeys, procure sustenance, and facilitate trade. In order to accomplish this, one must own a map of some kind and possess the skill to determine direction. Consequently, numerous navigation techniques have existed for millennia.

Early Pacific Islanders utilized stick charts, in which shells denoted islands, and curved sticks symbolized wave and current patterns surrounding the islands (Figure 2.1).

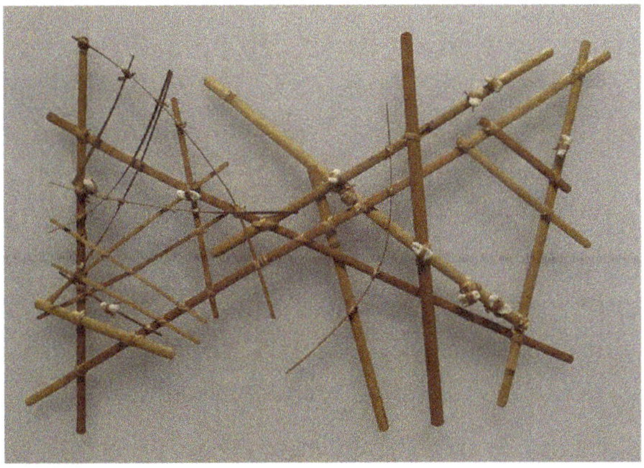

Figure 2.1 A Micronesian navigational chart from the Marshall Islands, made of wood, sennit fiber and cowrie shells (By Cullen328 – Own work, CC BY-SA 3.0, https://commons.wikimedia.org/w/index.php?curid=46844500).

The Phoenicians, over 4,000 years ago (c. 2000 B.C.E.), were the first Western culture to develop navigation at sea. Phoenician sailors achieved navigation by employing rudimentary charts and making observations of the sun and stars to ascertain directions. In 590 B.C.E., they conducted explorations of the Mediterranean and Red Seas, and successfully completed a circumnavigation of Africa.

Ptolemy was a Hellenistic polymath hailing from Alexandria, Egypt, who excelled in the fields of writing, mathematics, and science. In approximately 150 AD, he created cartographic representations of the known world, which included the positions of prominent towns. Additionally, he is credited with the earliest documented utilization of lines of latitude and longitude (Figure 2.2). The early coordinate system serves as the foundation for geolocation techniques that are still utilized today.

Figure 2.2 A 15th century world map based on Ptolemy's *Geography* (Public domain via Wikimedia Commons).

2.1 LATITUDE AND LONGITUDE

The coordinates of any location on Earth can be determined by the junction of its lines of latitude and longitude. Latitude is determined by measuring the angle between the equator, the Earth's center, and your location on the Earth's surface (Figure 2.1.1). Latitude is measured in degrees north or south of the equator (0°), with the poles being at a latitude of 90°. Consequently, the poles are designated as high latitude, while the equatorial region is classified as low latitude. Lines of identical latitude, known as parallels of latitude, are consistently equidistant from each other and do not converge. Nevertheless, the circles formed by the parallels of latitude do decrease in size as they approach the poles.

One degree of latitude is subdivided into 60 minutes ('). One unit of latitude is equivalent to one nautical mile, which is equal to 1.15 statute miles (1.85 kilometers). Each minute of latitude is further subdivided into 60 seconds ("). Traditionally, positions have been denoted using degrees, minutes, and seconds, for example, 36 degrees 15 minutes 32 seconds N. Nevertheless, due to the advent of contemporary digital technology, coordinates are now commonly represented as decimals, for instance, 36o 15.53′ N can be rendered as 36.2589o N. A valuable resource for converting coordinates between both formats can be accessed at: https://www.fcc.gov/media/radio/dms-decimal.

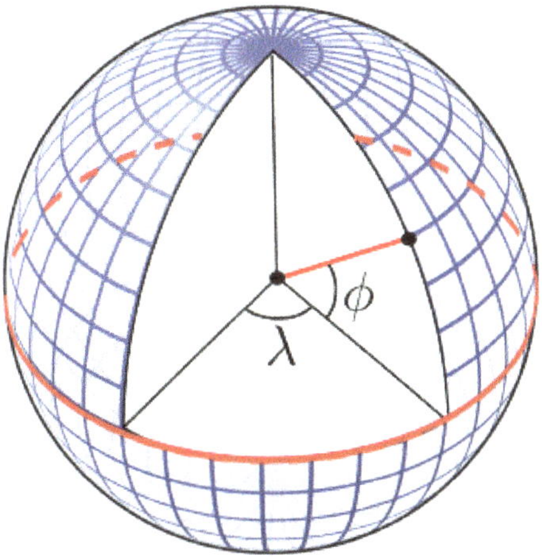

Figure 2.1.1 The latitude of a point on the Earth's surface is determined by the angle (ø) between the point and the equator, passing through Earth's **center (Peter Mercator [Public domain], via Wikimedia Commons).**

Latitude in the Northern Hemisphere can be ascertained by measuring the angle between the North Star (Polaris) and the horizon. The North Star is consistently positioned directly above the North Pole. At the North Pole, when a person gazes directly towards the horizon, the star will be positioned immediately above, forming a 90-degree angle. Therefore, the latitude at the North Pole is 90 degrees N. When observing from the equator and facing towards the north, the star and the horizon are aligned, resulting in an angle of 0o between them. Consequently, the equatorial latitude is also 0o. In the Northern Hemisphere, the latitude can be determined by measuring the angle between the horizon and the star at any location.

In ancient times, sailors utilized a device known as an astrolabe to determine this specific angle. Subsequently, the sextant was invented, enabling more precise measurements (Fig. 2.1.2).

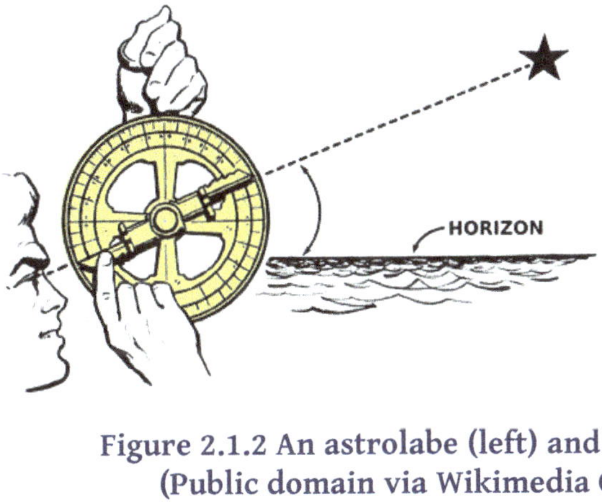

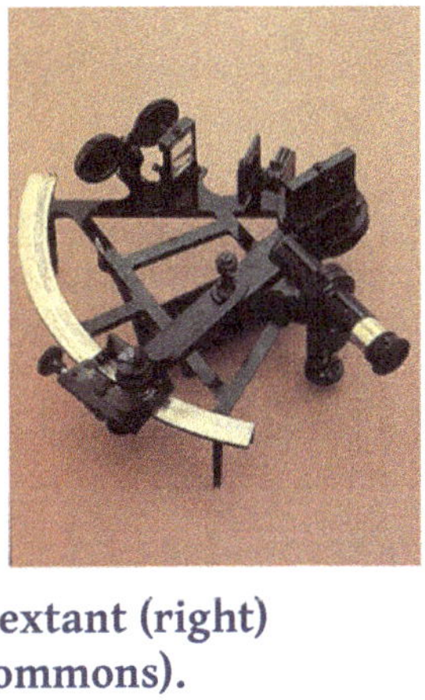

Figure 2.1.2 An astrolabe (left) and sextant (right) (Public domain via Wikimedia Commons).

There is no equivalent celestial object to the North Star in the Southern Hemisphere that can be effectively used to determine latitude. Nevertheless, the Southern Cross and Centaurus constellations can serve as reliable indicators for locating the south celestial pole. The intersection of the long axis of the Southern Cross and the line connecting the two brightest stars in Centaurus occurs at the south celestial pole.

Longitude is a measurement of the distance in the east or west direction from a reference point called the prime meridian (0o). The prime meridian is currently defined as the line that passes through Greenwich, England. However, throughout history, the prime meridian has been located in various places such as Rome, Copenhagen, Paris, Philadelphia, the Canary Islands, and Jerusalem. It is important to note that unlike the equator, the location of the prime meridian is somewhat arbitrary. The longitude is the angular distance, measured in degrees, between your location, the center of the Earth, and the prime meridian (Fig. 2.1.3). It indicates whether you are located to the east or west of the prime meridian.

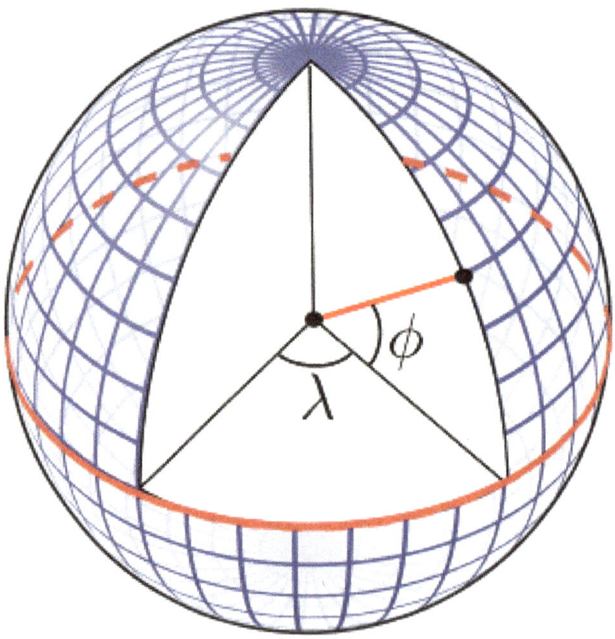

Figure 2.1.3 Longitude is determined as the angle (λ) between the prime meridian and your position (Peter Mercator [Public domain], via Wikimedia Commons).

As you travel in an eastward or westward direction from the prime meridian, you will ultimately reach the coordinates of 180 degrees east and west, which are located on the opposite side of the Earth from Greenwich. This location represents the International Date Line. Meridians of longitude, often known as great circles, are the lines of longitude. Unlike lines of latitude, which are parallel, all circles of longitude have the same length and converge as they approach the poles. Hence, whereas one minute of latitude always corresponds to one nautical mile, the length of one minute of longitude diminishes as one moves from the equator towards the poles, eventually reaching zero.

To measure longitude, one needs to have precise time at their current position, as well as the time at a distant point, such as a home port, at the exact same moment. Longitude can be calculated using the time difference. This is due to the fact that the Earth completes a full revolution of 360 degrees in a span of 24 hours. Within a span of one hour, the Earth completes a rotation equivalent to 1/24th of a full circle, which is 15 degrees. Hence, for every one-hour discrepancy in time between two places, there is a 15-degree variance in longitude.

Precise determinations of latitude utilizing the North Star have been conducted since at least the third century B.C.E. Due to the necessity of exact timekeeping, the measurement of longitude at sea became simply and accurately achievable only in the mid-18th century. Prior to that time, sailors frequently navigated by sailing either north or south to reach the required latitude, and then proceeded to travel east or west until they reached the intended longitude. The British government enacted the Longitude Act in 1714, recognizing the significance of resolving the longitude problem. They established a reward of £20,000 for anyone who could develop a technique to accurately measure longitude at sea with a precision of half a degree. Several unsuccessful methods were suggested, such as using astronomical observations, but it was John Harrison, a skilled clock manufacturer, who created a sequence of clocks that ultimately met the requirements. The initial iteration (H1) had a weight over 80 pounds, however the subsequent watches, H4 and H5, were compact enough to be grasped within the confines of a single hand. Paradoxically, despite meeting all the requirements, Harrison was never declared the victor of the longitude prize, and in truth, no official winner was ever established. By utilizing precise timekeeping devices, a ship has the capability to own two clocks: one synchronized with Greenwich time (or another designated home place), and another adjusted to local time, which may be reset daily by viewing the sun. The disparity in time between the two clocks can be utilized to determine longitude.

Currently, GPS (Global Positioning System) technology is employed to ascertain latitude and longitude. Furthermore, even the most compact smart phones and smart watches are equipped with GPS capabilities to compute position. GPS operates by means of a network of orbiting satellites that consistently broadcast signals comprising the time and their precise location. A GPS receiver collects signals from numerous satellites and uses triangulation to determine the position. The system requires a total of 24 satellites to be operational simultaneously. As of 2015, the system comprised approximately 32 functioning satellites, capable of providing a position with a precision of 9 meters (30 feet) or better.

2.2 MEASURING SPEED

Mariners needed to be aware of their speed at sea, in addition to their location and the water's depth. Early sailors utilized chip logs, which consisted of wooden boards affixed to elongated spooled ropes that included evenly spaced knots. The plank was discarded into the water, while the spool remained on the ship. Upon entering the water, the plank experienced resistance, causing it to remain relatively still as the connected line unwound. The pace of the line's unwinding indicated the speed at which the ship was moving away from the plank, and hence the ship's velocity. The velocity at which the line unwound was determined by tallying the quantity of knots that traversed the sailor's hands within a specific duration. Since it was originally measured as the number of knots per unit time, speed over water traditionally uses the unit of "knots" to indicate speed. One knot (kt) = 1 nautical mile per hour = 1.15 mph = 1.85 kph.

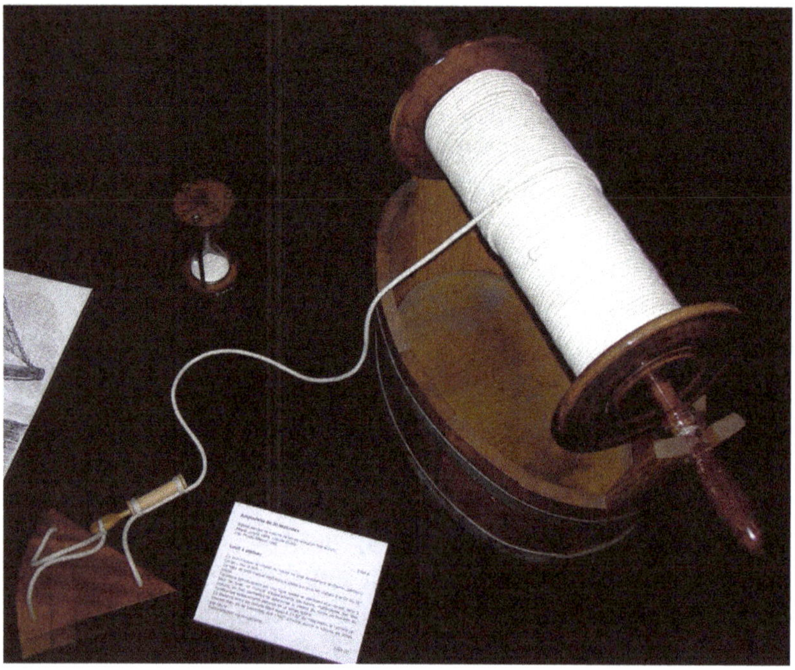

Figure 2.2.1 Chip log (Rémi Kaupp – Personal photograph taken in the Musée de la Marine, Paris, CC BY-SA 3.0, via Wikimedia Commons).

2.3 MAP PROJECTIONS

Studying oceanography necessitates the examination of maps, making it crucial to acknowledge the merits and limitations of the various map types one may come across. Representing a three-dimensional spherical object such as the Earth on a flat, two-dimensional map or chart is a challenging task. Hence, two-dimensional maps exhibit distortions when reflecting the Earth's actual area, direction, distance, and form. While globes are the only precise representation of these variables, they are not practical for field use and cannot be reproduced in a book. Due to these constraints, many map projections are employed to depict the Earth, based on the requirements of the presenter. The following are several prevalent map projections utilized in the field of oceanography.

Latitude and longitude in a Mercator projection are depicted as straight, parallel lines that connect at right angles (Figure 2.3.1). This projection is ideal for navigation as it accurately preserves directions. For instance, regardless of the location on the map, the north direction consistently points towards the top of the chart. As a result, Mercator projections have been the prevailing choice for navigational charts. An inherent limitation of this projection is the distortion of size and distance at high latitudes. The reason for this is that the spacing between lines of longitude decreases as you get closer to the poles, while on a Mercator projection, the spacing remains consistent. The poles, depicted as a singular point on a globe, are enlarged to match the circle of the equator. This overstates the distances, and consequently the area, at high latitudes. South America is almost nine times larger than Greenland. However, due to the distortion caused by the Mercator map projection, they appear to be the same size (Figure 2.3.1).

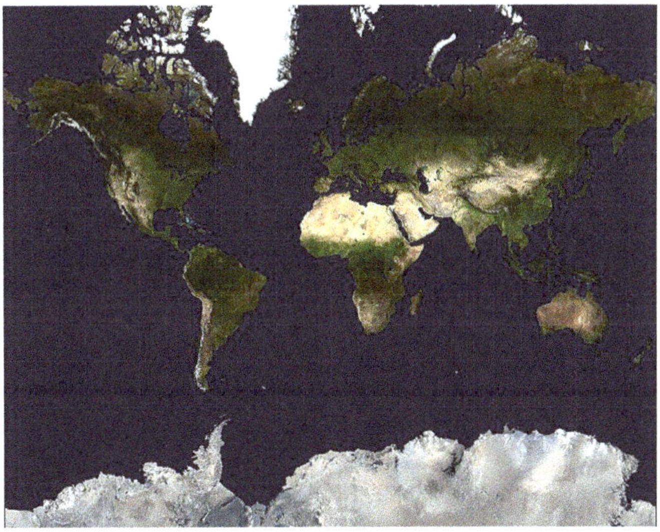

Figure 2.3.1 Mercator projection (By Mdf (Own work) [Public domain], via Wikimedia Commons).

The Goode homolosine projection is commonly employed for global representation (Figure 2.3.2). One benefit of this projection is that it minimizes the distortion of distance and area compared to the Mercator projection. However, there are also notable drawbacks. Clearly, the graphic below illustrates the issue of the oceans (including Greenland) being divided. Alternative iterations of this projection may preserve the oceans to some extent, but at the expense of disrupting the continents. It is impossible to preserve both the oceans and the continents in their entirety using this projection. The homolosine projection is unsuitable for navigation due to the varying orientations of the lines of longitude across different regions of the map.

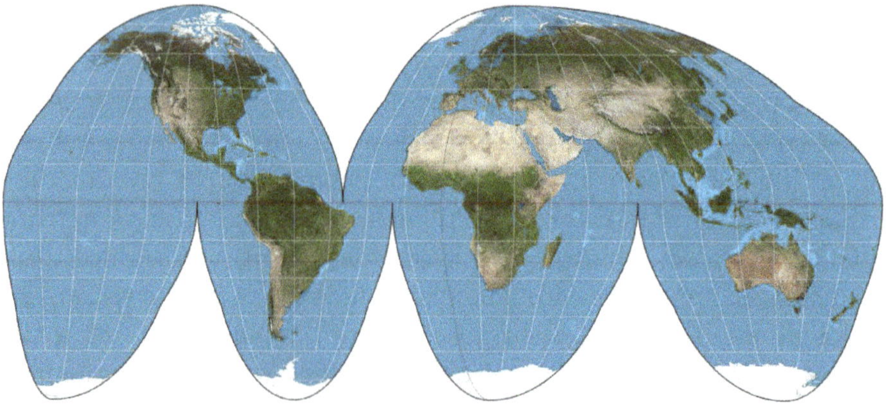

Figure 2.3.2 Goode homolosine projection (By Strebe (Own work) [CC BY-SA 3.0],via Wikimedia Commons).

The **Robinson planisphere projection** (Figure 2.3.3) keeps latitude horizontal, but shows some convergence of longitude. There is still some distortion, but not as much as in a Mercator. This projection is used mostly for data presentation.

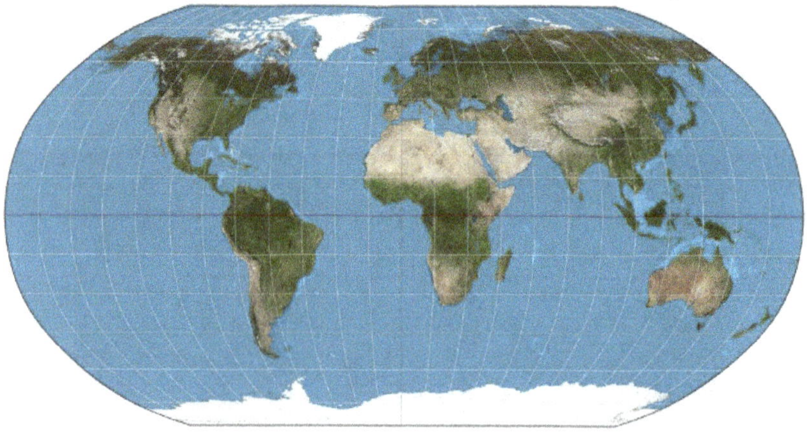

Figure 2.3.3 Robinson planisphere projection. (By Strebe (Own work) [CC BY-SA 3.0], via Wikimedia Commons).

In oceanography, our use of maps is not limited to viewing the Earth's surface; we also need to see what's at the bottom of the ocean. Other map types include **bathymetric** maps (Figure 2.3.4). These are similar to topographic maps for terrestrial locations, with lines connecting areas of equal depth. The closer together the lines, the steeper the feature. In the example below, the steep continental slope is represented by the high density of depth contours as the colors transition from light blue to dark blue. The well-spaced dark blue lines in the bottom right of the figure represent the relatively flat deep seafloor.

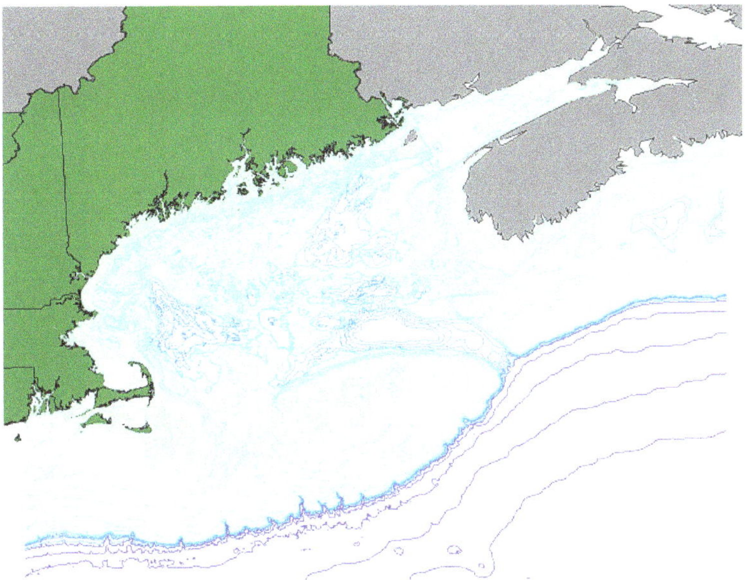

Figure 2.3.4 Bathymetric map of the Gulf of Maine (USGS).

Physiographic maps display bathymetry data in the form of a three-dimensional relief map, illustrating the characteristics of the ocean (Figure 2.3.5). It is crucial to acknowledge that they often exhibit substantial vertical exaggeration. In the given example, the coastline spans several hundred kilometers, although the difference in depth from the continental shelf to the seafloor is only a few kilometers.

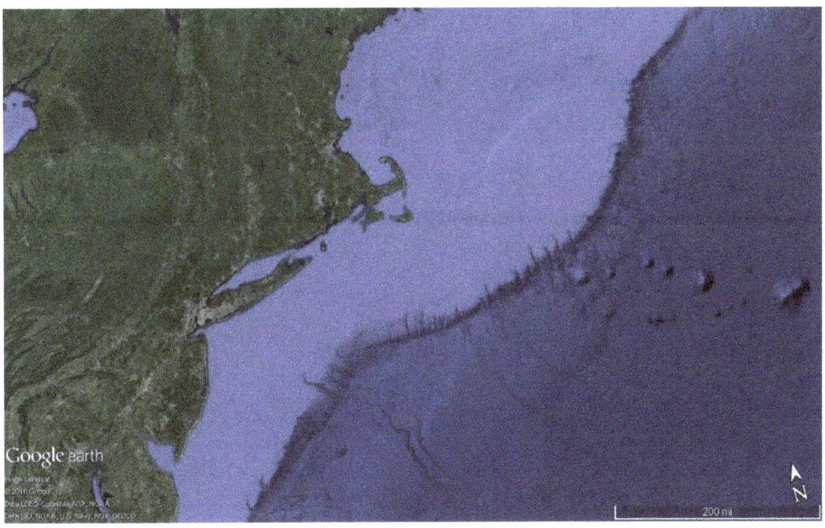

Figure 2.3.5 Physiographic map of the southern New England coast (Google Earth, Map Data: SIO, NOAA, US Navy, NGA, GEBCO).

3 Chapter

THE GENESIS AND COMPOSITION OF EARTH

3.1 ORIGIN OF EARTH AND THE SOLAR SYSTEM

*Modified from Karla Panchuk in "Physical Geology" by Steven Earle**

Based on the Big Bang theory, the universe abruptly came into existence 13.77 billion years ago (Figure 3.1.1). The Big Bang is commonly characterized as an explosion, yet envisioning it as a colossal fireball is not precise. The Big Bang entailed an abrupt proliferation of matter, energy, and space originating from a solitary point. The Hollywood explosion that is typically imagined entails the expansion of matter and energy within space. However, during the big bang, space itself was actually formed.

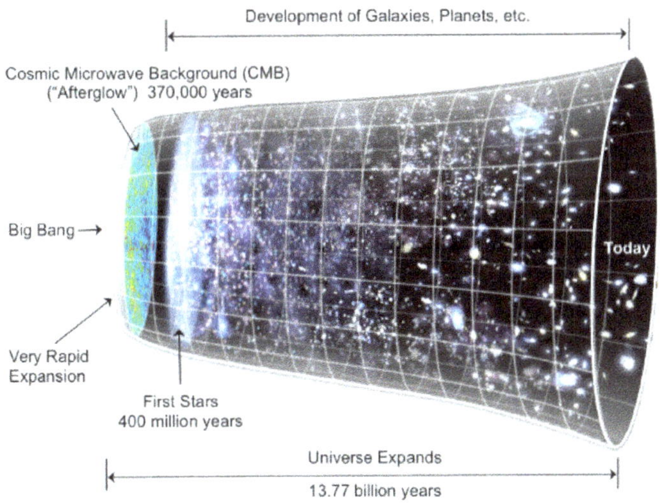

Figure 3.1.1 The Big Bang and development of the universe (Steven Earle, "Physical Geology").

During the initial stages of the Big Bang, the universe was in a state of extreme heat and density, consisting of particles smaller than atoms. However, as the cosmos expanded, it also underwent a cooling process. Over time, a portion of the particles collided and adhered to one another. The collisions resulted in the formation of hydrogen and helium, which are the most abundant elements in the universe, as well as a minor quantity of lithium. The force of gravity led to the aggregation of clouds containing these primitive elements, resulting in the formation of stars. It was within these stars that the synthesis of heavier elements took place.

The formation of our solar system commenced approximately 5 billion years ago, which was approximately 8.7 billion years subsequent to the occurrence of the Big Bang. A solar system is comprised of a group of celestial bodies that revolve around one or more central stars. Every solar system begins in an identical manner. Their origin lies within a nebula, which is a mixture of gas and dust. Nebulae are visually stunning celestial objects captured in space photography, showcasing bright hues resulting from the presence of gasses and dust, as well as dazzling flickering caused by the numerous stars that have originated within them (Figure 3.1.2). The gas mostly comprises hydrogen and helium, whereas the dust is composed of minute mineral grains, ice crystals, and organic particles.

Figure 3.1.2 Photograph of a nebula. The Pillars of Creation within the Eagle Nebula viewed in visible light (left) and near infrared light (right). Near infrared light captures heat from stars, and allows us to view stars that would otherwise be hidden by dust. This is why the picture on the right appears to have more stars than the picture on the left [NASA, ESA, and the Hubble Heritage Team (STScI/AURA) http://bit.ly/1Dm2X5a].

The formation of a solar system commences when a relatively small region within a nebula (little in comparison to the vastness of the universe) undergoes gravitational collapse. The precise initiation of this phenomenon remains uncertain, although it is possible that it is instigated by the tumultuous conduct of neighboring stars as they undergo their evolutionary processes. The energy and substance emitted by these stars have the potential to compress the gas and dust in the surrounding regions of the nebula. After being triggered, the collapse of gas and dust inside that specific area persists due to two factors. One explanation for this is that the force of gravity attracts gas molecules and dust particles, causing them to come together. However, at the initial stages, the size of these particles is minuscule, resulting in a weak gravitational pull between them. How do they converge? Dust initially gathers in loose agglomerations due to the presence of static electricity, similar to the formation of dust bunnies beneath one's bed. As the compact region within a nebula undergoes condensation, the process of star formation commences as matter is pulled towards the center of the region. The residual dust and gas then settle into a rotating disk surrounding the newly formed star. A protoplanetary disk is the location where planets eventually develop. Figure 3.1.3 displays an artist's depiction of a protoplanetary disk in the upper left, whereas the upper right image presents an authentic protoplanetary disk encircling the star HL Tauri. Observe the presence of the opaque bands within the protoplanetary disk. These gaps represent the initial stages of planet formation. The presence of rings is attributed to the accumulation of dust and gas in the orbits of developing planets. An comparison can be drawn between the black rings in question and the gaps in Saturn's rings, as depicted in Figure 3.1.3 (lower left). These gaps in Saturn's rings are known to include moons, as shown in Figure 3.1.3 (lower right).

Figure 3.1.3 Protoplanetary disks and Saturn's rings. Upper left: An artists impression of a protoplanetary disk containing gas and dust, surrounding a new star. [NASA/ JPL-Caltech, http://1.usa.gov/1E5tFJR] Upper right: A photograph of the protoplanetary disk surrounding HL Tauri. The dark rings within the disk are thought to be gaps where newly forming planets are sweeping up dust and gas. [ALMA (ESO/NAOJ/NRAO) http://bit.ly/1KNCq0e]. Lower left: A photograph of Saturn showing similar gaps within its rings. The bright spot at the bottom is an aurora, similar to the northern lights on Earth. [NASA, ESA, J. Clarke (Boston University), and Z. Levay (STScI) http://bit.ly/1IfSCX5] Lower right: a close-up view of a gap in Saturn's rings showing a small moon as a white dot. [NASA/JPL/ Space Science Institute, http://1.usa.gov/1g2EeYw].

Typically, planets can be categorized into three groups according to their composition (Fig. 3.1.4). Terrestrial planets, such as Earth, Mercury, Venus, and Mars, possess a metallic core enveloped by rock. Jovian planets, often known as gas giants, are celestial bodies such as Jupiter and Saturn that are primarily composed of hydrogen and helium. Ice giants, like Uranus and Neptune, are predominantly composed of water ice, methane (CH4) ice, and ammonia (NH3) ice, with rocky cores. Uranus and Neptune, although often classified as gas giants alongside Jupiter and Saturn, differ significantly from them.

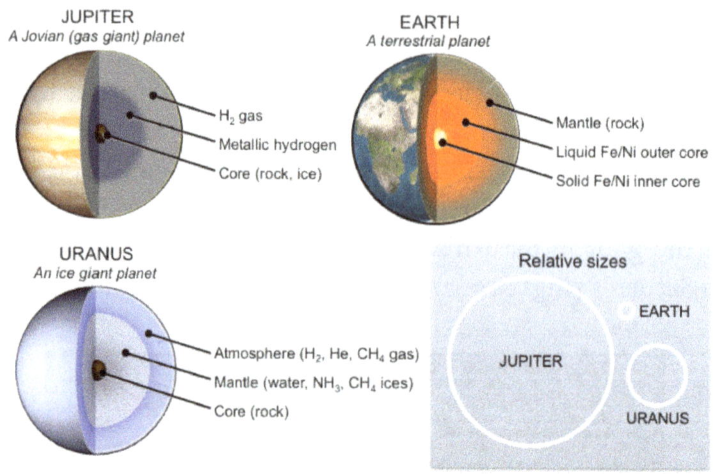

Figure 3.1.4 Three types of planets. Jovian (or gas giant) planets such as Jupiter consist mostly of hydrogen and helium. They are the largest of the three types. Ice giant planets such as Uranus are the next largest. They contain water, ammonia, and methane ice. Terrestrial planets such as Earth are the smallest, and they have metal cores covered by rocky mantles. [KP, after public domain images by Francesco A, Wolfman SF (http://bit.ly/1eP75P4), and NASA (http://1.usa.gov/1gFVsf6, http://1.usa.gov/1M89jI3)].

The arrangement of these three sorts of planets inside our solar system is not haphazard. Instead, they follow a regular pattern, with the terrestrial planets positioned closest to the sun, followed by the Jovian planets, and finally the ice giants. The frost line, often known as the snow line, is one of the reasons for this design. The frost line acted as a boundary within the protoplanetary disk, dividing the region closer to the sun, where the temperature was too high for anything other than silicate minerals and metal to solidify, from the outer region farther from the sun, where the temperature was low enough for ice to develop. Consequently, the items that originated in the inner region of the protoplanetary disk are primarily composed of rock and metal, whereas the objects that formed in the outer region are predominantly composed of gas and ice. The youthful sun emitted intense solar winds composed of energetic particles, which facilitated the movement of lighter molecules towards the outer region of the protoplanetary disk.

The celestial bodies in our solar system were created through the process of accretion. At the beginning of this process, mineral and rock particles gathered together in soft clusters due to static electricity. As the mass of the clumps increased, the force of gravity became increasingly significant, causing material to be pulled from greater distances and resulting in the growth of these solid masses into progressively larger bodies. Over time, the particles grew in size until their gravitational force became powerful enough to retain gas molecules, which are inherently lightweight.Our Earth formed though this process of accretion about 4.6 billion years ago. The early Earth was very hot and had a molten, fluid composition, with lost of geological and volcanic activity on the surface. The Earth's heat came from a variety of processes:

- Heat came from the decay of radioactive elements within the Earth, specifically the decay of 235U, 238U, 40K, and 232Th, which are primarily present in the mantle. The total heat produced that way has been decreasing over time (because these isotopes are getting used up), and is now roughly 25% of what it was when Earth formed. This means that Earth's interior is slowly becoming cooler.
- Heat came from the thermal energy already contained within the objects that accreted to form the Earth.
- Heat came from collisions. When objects hit Earth, some of the energy from their motion went into deforming Earth, and some of it was transformed into heat. (The very worst collision that Earth experienced was with a planet named Theia, which was approximately the size of Mars. Not long after Earth formed, Theia struck Earth. When Theia slammed into Earth, Theia's metal core merged with Earth's core, and debris from the outer silicate layers was cast into space, forming a ring of rubble around Earth. The material within the ring coalesced into a new body in orbit around Earth, giving us our moon. Remarkably, the debris may have coalesced in 10 years or fewer! This scenario for the formation of the moon is called the **giant impact hypothesis**.)
- As Earth became larger, its gravitational force became stronger. This increased Earth's ability to draw objects to it, but it also caused the material making Earth to be compressed, rather like Earth giving itself a giant gravitational hug. Compression causes materials to heat up.

Heating had a very important consequence for Earth's structure. As Earth grew, it collected a mixture of silicate mineral grains as well as iron and nickel. These materials were scattered throughout Earth. However, this situation altered as the Earth started to experience a rise in temperature, reaching such high levels that it caused the melting of both silicate minerals and metals. The metal melt had a significantly higher density compared to the silicate material melt. Consequently, the metal melt descended towards the Earth's center, ultimately becoming its core. Conversely, the silicate melt ascended towards the surface, giving rise to the Earth's crust and mantle. To put it simply, the Earth separated into its constituent parts. Differentiation refers to the process of segregating silicate minerals and metals into distinct layers, with silicate minerals forming the outer rocky layer and metals forming the inner metallic core. Due to the force of gravity, Earth has been shaped into a nearly perfect sphere with a radius of 6371 km and a circumference of around 40,000 km. Nevertheless, the Earth is not a perfectly spherical shape due to the effect of its rotation. This generates a bulge at the equator, resulting in the Earth's circumference being 21 km (0.3%) larger at the equator compared to the distance between the poles. Therefore, it can be described as a "oblate spheroid" from a technical standpoint.

An inventory of Earth's composition reveals that four elements, namely oxygen, magnesium, silicon, and iron, account for 95% of its mass. The majority of the remaining 5% is derived from aluminum, calcium, nickel, hydrogen, and sulfur. While the Big Bang is responsible for the creation of hydrogen, helium, and lithium, the origins of the remaining elements are still a subject of inquiry. The explanation lies in the fact that the remaining components were synthesized through stellar processes. High temperatures and intense pressure within stars facilitate the collision and fusion of smaller atoms, resulting in the formation of bigger atoms. When hydrogen atoms collide and merge, they undergo fusion and produce helium. Significant quantities of energy are liberated during the fusion of certain atoms, and it is this energy that is responsible for the luminosity of stars.

Stars of greater magnitude are required to produce elements of such high atomic weight as iron and nickel. The Sun, a typical star, will reach the end of its lifespan once it depletes its hydrogen fuel and undergoes fusion to produce helium, beryllium, carbon, nitrogen, oxygen, and fluorine. The process of atom production will cease, and the object will gradually decrease in temperature and expand until its central region reaches the orbit of Mars. Conversely, massive stars conclude their existence in a remarkable manner by undergoing a supernova explosion, thereby releasing newly created atoms, including elements that are more massive than iron, into the surrounding space. It required numerous stellar generations to produce heavier elements and disperse them into space until there was a sufficient abundance to enable the formation of planets such as Earth.

3.2 STRUCTURE OF EARTH

*Portions modified from "Physical Geology" by Steven Earle**

In the preceding section, we acquired knowledge that the components present on the early Earth were segregated by the process of differentiation. This involved the sinking of denser materials such as iron and nickel towards the center, while lighter materials like oxygen, silicon, and magnesium remained closer to the surface. Consequently, the Earth is structured with layers of varying composition and density that progressively increase as one moves from the surface towards the core. (Figure 3.2.1).

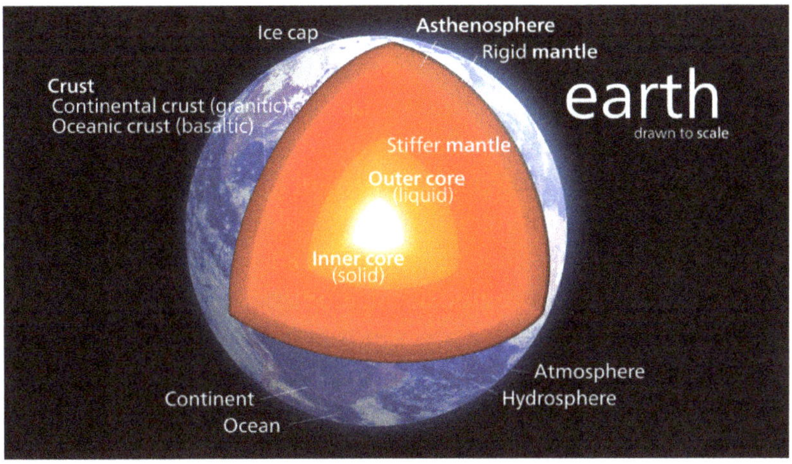

Figure 3.2.1 Interior structure of Earth (By Kelvinsong (Own work) [CC BY-SA 3.0], via Wikimedia Commons).

The traditional view based on chemical composition recognizes four distinct layers:

The **inner core** lies at the center of the Earth, and is about 1200 km thick. It is composed primarily of iron alloys and nickel, with about 10% comprised of oxygen, sulfur or hydrogen. The temperature in the inner core is about 6000 °C (10,800 °F), which is roughly the temperature of the surface of the sun (section 3.1 explains the sources of this intense heat). Despite the high temperature that should melt these metals, the extreme pressure (from literally the weight of the world) keeps the inner core in the solid phase. The solid metals also make the inner core very dense, at about 17 g/cm^3, giving the inner core about one-third of the Earth's total mass.

The **outer core** sits outside of the inner core. It has the same composition as the inner core, but it exists as a fluid, rather than a solid. The temperature is 4000-6000 °C, and the metals remain in the liquid state because the pressure is not as great as in the inner core. It is the movement of the fluid iron in the outer core that creates Earth's magnetic field (see section 4.2). The outer core is 2300 km thick, and has a density of 12 g/cm^3.

The **mantle** extends from the outer core to just under Earth's surface. It is 2900 km thick, and contains about 80% of the Earth's volume. The mantle consists of iron and magnesium silicates and magnesium oxides, so it is more similar to the rocks of Earth's surface than to the materials in the core. The mantle has a density of 4.5 g/cm³, and temperatures in the range of 1000-1500 °C. The uppermost layer of the mantle is more rigid, while the deeper regions are fluid, and it is the motion of fluid materials in the mantle that is responsible for plate tectonics (see section 4.3). Magma that rises to the surface through volcanoes originates in the mantle.

The outermost layer is the **crust**, which forms the solid, rocky surface of the Earth. The crust averages 15-20 km thick, but in some places, such as under mountains, the crust can reach thicknesses of up to 100 km. There are two main types of crust; **continental crust** and **oceanic crust** that differ in a number of ways. Continental crust is thicker than oceanic crust, averaging 20-70 km thick, compared to 5-10 km for oceanic crust. Continental crust is also older than oceanic crust; the oldest rocks in continental crust are about 4.4 billion years old, while the oldest oceanic crust only goes back about 180 million years. Finally, the two types of crust differ in their composition. Continental crust is made largely of granite, while oceanic crust is mostly composed of basalt. Basalt contains a higher proportion of heavier elements like iron and magnesium, making oceanic crust denser than continental crust (3 g/cm³ vs. 2.7 g/cm³.) Granite also contains larger crystals than basalt, as underground or surface continental magmas can cool slowly, which allows time for crystal structures to form before the rocks solidify. Basalt also forms from cooling magmas, but exposure to air or water can cause them to cool faster, which does not allow time for large crystals to form.

Based on physical characteristics, we can also divide the outermost layers of Earth into the **lithosphere** and **asthenosphere**. The lithosphere consists of the crust and the cool, rigid, outer 80-100 km of the mantle. The crust and outer mantle moves together as a unit, so they are combined together into the lithosphere. The asthenosphere lies below the lithosphere, from about 100-200 km to about 670 km deep. It includes the more "plastic" softer region of the mantle, where fluid movements can occur. The solid lithosphere is thus floating on the fluid asthenosphere.

Isostasy

To help explain how the lithosphere is floating on the asthenosphere, we need to examine the concept of **isostasy**. Isostasy refers to the way a solid will float on a fluid. The relationship between the crust and the mantle is illustrated in Figure 3.2.2. On the right is an example of a non-isostatic relationship between a raft and solid concrete. It's possible to load the raft up with lots of people, and it still won't sink into the concrete. On the left, the relationship is an isostatic one between two different rafts and a swimming pool full of peanut butter. With only one person on board, the raft floats high in the peanut butter, but with three people, it sinks dangerously low. We're using peanut butter here, rather than water, because its viscosity more closely represents the relationship between the crust and the mantle. Although it has about the same density as water, peanut butter is much more viscous (stiff), and so although the three-person raft will sink into the peanut butter, it will do so quite slowly.

Figure 3.2.2 Demonstrating isostasy (Steven Earle, "Physical Geology").

The relationship of Earth's crust to the mantle is similar to the relationship of the rafts to the peanut butter. The raft with one person on it floats comfortably high. Even with three people on it the raft is less dense than the peanut butter, so it floats, but it floats uncomfortably low for those three people. The crust, with an average density of around 2.6 grams per cubic centimeter (g/cm^3), is less dense than the mantle (average density of approximately 3.4 g/cm^3 near the surface, but more than that at depth), and so it is floating on the "plastic" mantle. When more weight is added to the crust, through the process of mountain building, it slowly sinks deeper into the mantle and the mantle material that was there is pushed aside (Figure 3.2.3, left). When that weight is removed by erosion over tens of millions of years, the crust rebounds and the mantle rock flows back (Figure 3.2.3, right).

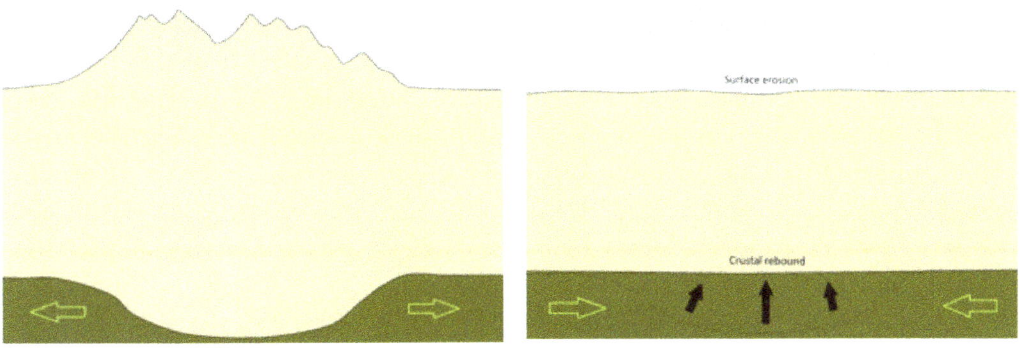

Figure 3.2.3 Isostatic rebound when mass is removed from the crust (Steven Earle, "Physical Geology").

The crust and mantle respond in a similar way to glaciation. Thick accumulations of glacial ice add weight to the crust, and as the mantle beneath is squeezed to the sides, the crust subsides. When the ice eventually melts, the crust and mantle will slowly rebound, but full rebound will likely take more than 10,000 years. Large parts of Canada are still rebounding as a result of the loss of glacial ice over the past 12,000 years, and as shown in Figure 3.2.4, other parts of the world are also experiencing isostatic rebound. The highest rate of uplift is in within a large area to the west of Hudson Bay, which is where the Laurentide Ice Sheet was the thickest (over 3,000 m). Ice finally left this region around 8,000 years ago, and the crust is currently rebounding at a rate of nearly 2 cm/year.

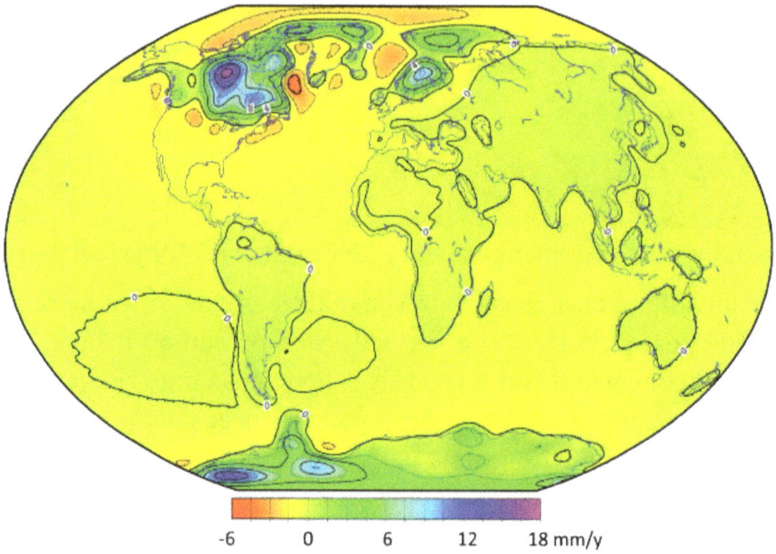

Figure 3.2.4 Global rates of isostatic adjustment (Steven Earle, "Physical Geology").

Since continental crust is thicker than oceanic crust, it will float higher and extend deeper into the mantle than oceanic crust. The boundary between the crust and the mantle is known as the **Mohorovičić discontinuity** (or **Moho**; see section 3.3). Crust is thickest where there are mountains, so the Moho will be deeper under mountains than under the oceanic crust. Since oceanic crust is also denser than continental crust, it floats lower on the mantle. Since the oceanic crust lies lower than the continental crust, and since water flows downhill to reach the lowest point, this explains why water has accumulated over the oceanic crust to form the oceans.

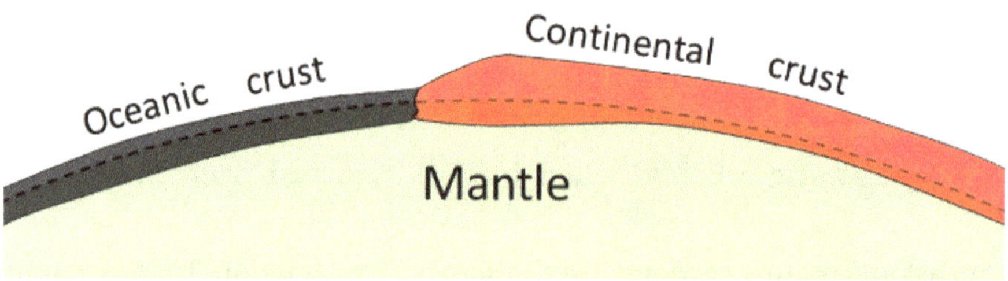

Figure 3.2.5 Thinner, denser oceanic crust floats lower on the mantle than thicker, less dense continental crust (Steven Earle, "Physical Geology").

3.3 DETERMINING THE STRUCTURE OF EARTH

*Modified from "Physical Geology" by Steven Earle**

The previous section described the properties and composition of Earth's interior, which begs the question: how can we know what conditions are like deep in the

Earth? It's easy to sample the crust through drilling, and mantle material often comes to the surface as magma, but the farthest we have been able to drill into the crust so far is only about 12 km; this for a planet with a radius of 6370 km! So to understand the composition and structure of the Earth's deep interior, we need to use indirect methods such as seismology.

Seismology is the study of vibrations within the Earth. These vibrations are caused by various events, including earthquakes, extraterrestrial impacts, explosions, storm waves hitting the shore, and tidal effects. Of course, seismic techniques have been most widely applied to the detection and study of earthquakes, but there are many other applications, and arguably seismic waves provide the most important information that we have concerning Earth's interior. Before going any deeper into Earth, however, we need to take a look at the properties of seismic waves. The types of waves that are useful for understanding Earth's interior are called **body waves**, meaning that, unlike the surface waves on the ocean, they are transmitted through Earth materials.

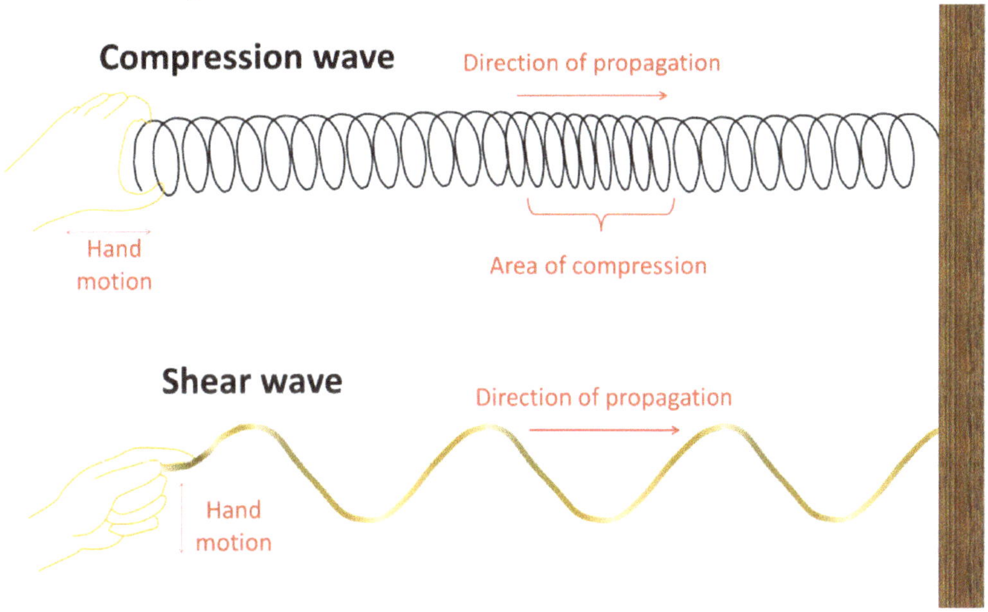

Figure 3.3.1 Representations of a compression wave (P-wave, top) and a shear wave (S-wave, bottom) (Steven Earle, "Physical Geology").

Compression waves and shear waves travel very quickly through geological materials. As shown in Figure 3.3.2, typical P-wave velocities are between 0.5 km/s and 2.5 km/s in unconsolidated sediments, and between 3.0 km/s and 6.5 km/s in solid crustal rocks. Of the common rocks of the crust, velocities are greatest in basalt and granite. S-waves are slower than P-waves, with velocities between 0.1 km/s and 0.8 km/s in soft sediments, and between 1.5 km/s and 3.8 km/s in solid rocks.

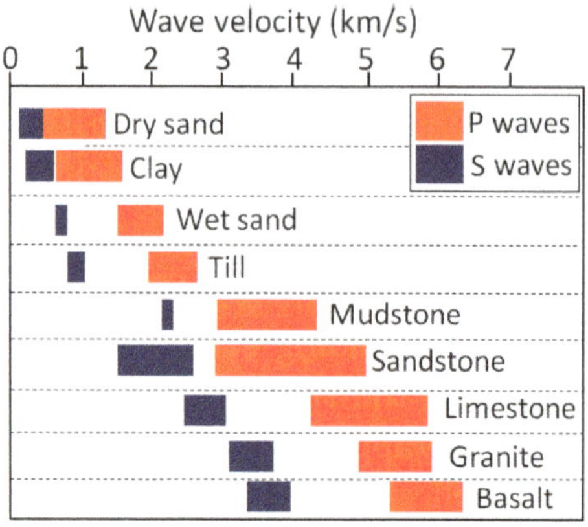

Figure 3.3.2 Typical velocities of P waves (red) and S waves (blue) in sediments and in solid crustal rocks (Steven Earle, "Physical Geology").

Mantle rock is generally denser and stronger than crustal rock and both P- and S-waves travel faster through the mantle than they do through the crust. Moreover, seismic-wave velocities are related to how tightly compressed a rock is, and the level of compression increases dramatically with depth. Finally, seismic waves are affected by the phase state of rock. They are slowed if there is any degree of melting in the rock. If the material is completely liquid, P-waves are slowed dramatically and S-waves are stopped altogether.

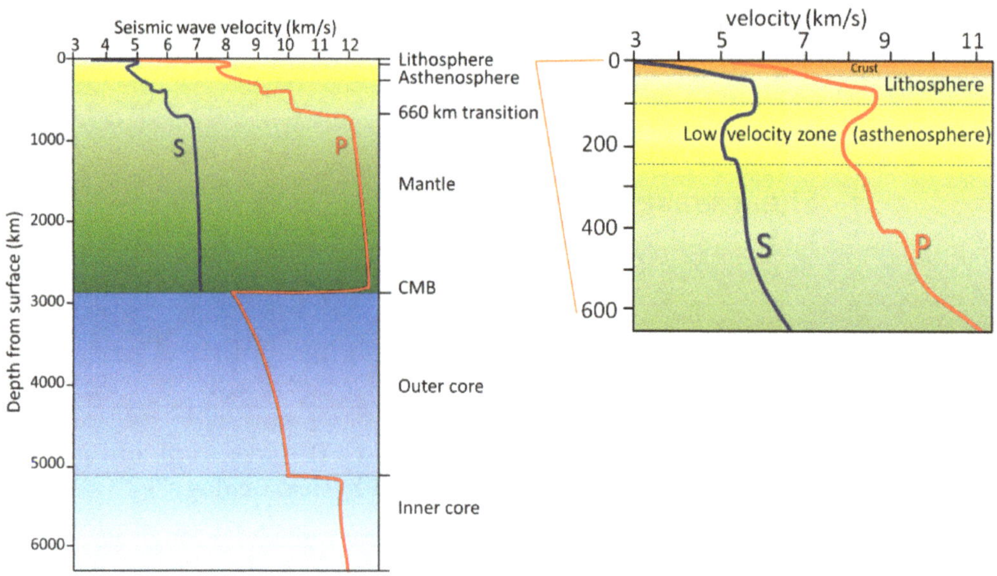

Figure 3.3.3 Wave velocities through the different layers of the Earth (left). Enhanced view of wave velocities in the crust and upper mantle (right) (Steven Earle, "Physical Geology").

Accurate seismometers have been used for earthquake studies since the late 1800s, and systematic use of seismic data to understand Earth's interior started in the early 1900s. The rate of change of seismic waves with depth in the Earth (Fig. 3.3.3) has been determined over the past several decades by analyzing seismic signals from large earthquakes at seismic stations around the world. Small differences in arrival time of signals at different locations have been interpreted to show that:

- Velocities are greater in mantle rock than in the crust.
- Velocities generally increase with pressure, and therefore with depth.
- Velocities slow in the area between 100 km and 250 km depth (called the "low-velocity zone"; equivalent to the asthenosphere).
- Velocities increase dramatically at 660 km depth (because of a mineralogical transition).
- Velocities slow in the region just above the core-mantle boundary (the D" layer or "ultra-low-velocity zone").
- S-waves do not pass through the outer part of the core.
- P-wave velocities increase dramatically at the boundary between the liquid outer core and the solid inner core.

One of the first discoveries about Earth's interior made through seismology was in the early 1900s when Croatian seismologist Andrija Mohorovičić (pronounced *Moho-ro-vi-chich*) realized that at certain distances from an earthquake, two separate sets of seismic waves arrived at a seismic station within a few seconds of each other. He reasoned that the waves that went down into the mantle, traveled through the mantle, and then were bent upward back into the crust, reached the seismic station first because although they had farther to go, they traveled faster through mantle rock (as shown in Figure 3.3.4). The boundary between the crust and the mantle is known as the **Mohorovičić discontinuity** (or **Moho**). Its depth is between 60 km and 80 km beneath major mountain ranges, around 30 km to 50 km beneath most of the continental crust, and between 5 km and 10 km beneath the oceanic crust.

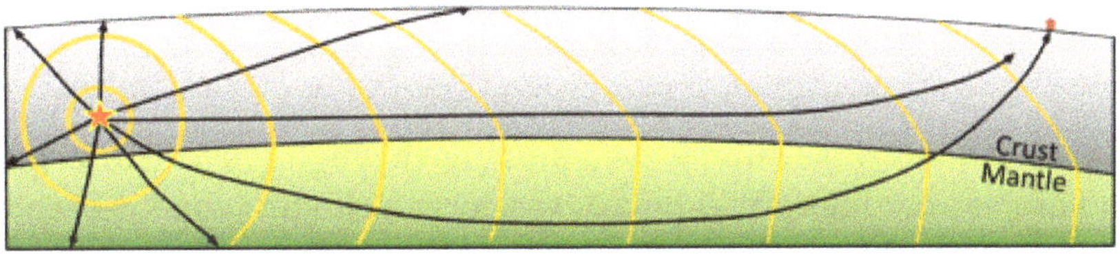

Figure 3.3.4 Depiction of seismic waves emanating from an earthquake (red star). Some waves travel through the crust to the seismic station (at about 6 km/s), while others go down into the mantle (where they travel at around 8 km/s) and are bent upward toward the surface, reaching the station before the ones that traveled only through the crust (Steven Earle, "Physical Geology").

Figure 3.3.5 provides a concise summary of our present comprehension of the patterns of seismic wave propagation throughout the Earth. Due to the progressive rise in density as depth increases, all waves are bent towards the region of lower density and slower velocity as they pass through uniform sections of the Earth. As a result, they tend to curve away from the center and towards the surface. Waves undergo refraction when they encounter barriers within the Earth, such as the Moho, the core-mantle boundary (CMB), and the outer-core/inner-core boundary. S-waves are unable to propagate through liquids and are halted at the core-mantle boundary (CMB). As a result, there exists a region on the opposite side of the Earth from a seismic source where S-waves are absent, known as the S-wave shadow. The seismic source is located at an angular distance of 103° from the shadow zone on both sides. Therefore, the total angular distance of the shadow zone is 154°. This information can be utilized to deduce the depth to the Cosmic Microwave Background (CMB). P-waves are capable of propagating across liquids, allowing them to penetrate the liquid component of the core. Due to refraction occurring at the CMB, waves passing through the core are deflected away from the surface, resulting in the formation of a P-wave shadow zone on both sides, spanning from 103° to 150°. This data can be utilized to discern the disparities between the internal and external regions of the core.

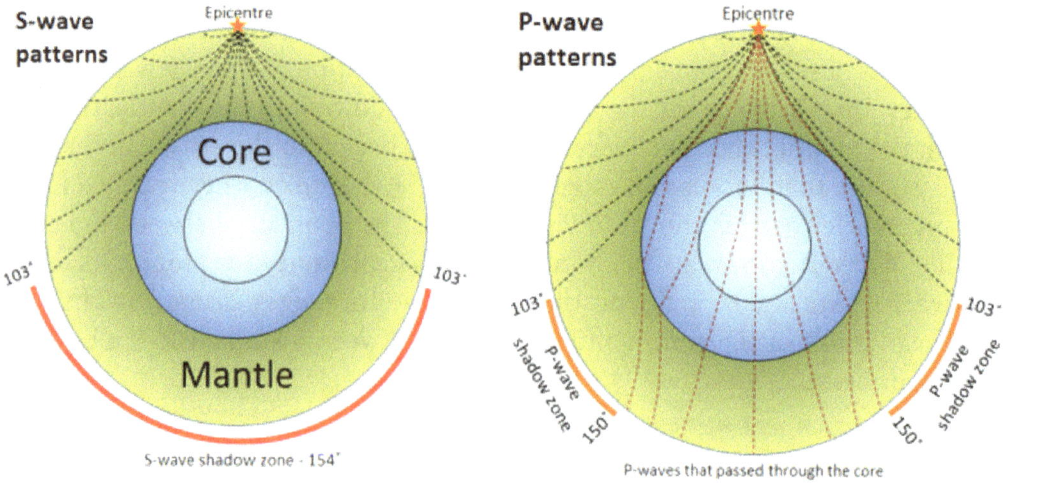

Figure 3.3.5 Patterns of seismic waves moving through Earth's interior. Since S waves do not pass through the liquid outer core, a shadow zone is created on the opposite side from the original disturbance (Steven Earle, "Physical Geology").

Using data from many seismometers and hundreds of earthquakes, it is possible to create a two- or three-dimensional image of the seismic properties of part of the mantle. This technique is known as seismic tomography, and an example of the result is shown in Figure 3.3.6.

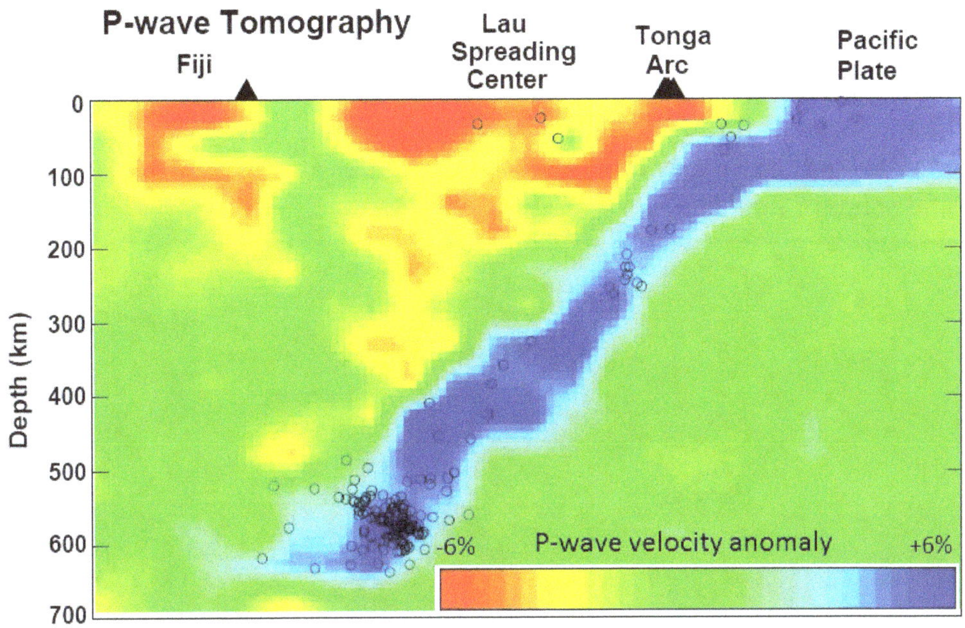

Figure 3.3.6 Seismic tomography image showing the Pacific Plate (blue) ubducting beneath Tonga (Steven Earle, "Physical Geology").

The Pacific Plate subducts beneath Tonga and appears in Figure 3.3.6 as a 100 km thick slab of cold (blue-colored) oceanic crust that has pushed down into the surrounding hot mantle. The cold rock is more rigid than the surrounding hot mantle rock, so it is characterized by slightly faster seismic velocities. There is volcanism in the Lau spreading center and also in the Fiji area, and the warm rock in these areas has slower seismic velocities (yellow and red colors).

Seismic waves offer insights into the composition and characteristics of the Earth's interior, but what about its additional attributes? Regarding the composition, multiple lines of evidence indicate a mostly iron and nickel core. Wave characteristics indicate that the core is made up of an element with an atomic number about 25 (iron has an atomic number of 26). Except for iron, all the other elements with a similar atomic number to 25 are too scarce to constitute the core. If the Earth originated from the accumulation of smaller celestial objects like meteorites, it would be reasonable to anticipate that the composition of the Earth would resemble that of meteorites. Meteorites consist primarily of iron and nickel, albeit in higher concentrations compared to the Earth's crust. These findings indicate that the majority of the dense iron and nickel present in the meteorites likely descended to the Earth's core during the planet's formation. Nevertheless, the central region lacks the necessary density to consist only of iron and nickel; it falls approximately 10% below the expected density for such a composition. Scientists hypothesize that the core is comprised of approximately

10% sulfur, oxygen, and hydrogen. Ultimately, the presence of iron in the outer core is a necessary condition for the Earth's magnetic field to originate from the fluid outer core. Regarding the temperatures, we can determine the melting points of these substances throughout the various pressures they encounter in the Earth's interior. From this, we may deduce the temperatures at which these elements can exist as solids or liquids.

Chapter 4 THE STUDY OF PLATE TECTONICS AND MARINE GEOLOGY

4.1 ALFRED WEGENER AND THE THEORY OF PLATE TECTONICS

*Modified from "Physical Geology" by Steven Earle**

Upon examining a map of Earth, one may observe that certain continents appear to be capable of fitting together. Francis Bacon, in the 17th century, made an early observation of this occurrence, noting the resemblances between the Atlantic coasts of Africa, and North and South America. The continents' current alignment is a result of their previous connection and subsequent separation, known as continental drift. Nevertheless, it has been established that it is not solely the continents that undergo movement, hence the most accurate phrase to describe this phenomenon is plate tectonics. The credit for the origin of this idea can be attributed to Alfred Wegener (Figure 4.1.1).

Figure 4.1.1 Prof. Dr. Alfred Wegener, ca. 1924-1930 (Public domain, via Wikimedia Commons).

Alfred Wegener (1880-1930) obtained a doctorate in astronomy from the University of Berlin in 1904. However, his primary academic focus and career were centered around geophysics and meteorology. In 1911, he came upon a scholarly journal that contained a description of the presence of corresponding Permian-aged land fossils in different regions of South America, Africa, India, Antarctica, and Australia (Figure 4.1.2). Wegener deduced that the presence of these fossils could only be explained if these continents were formerly connected. Moreover, several transcontinental regions exhibit comparable fossil records until around 150 million years ago, after which they start to diverge. This indicates that these regions eventually divided, leading to distinct paths of speciation on each continent. Wegener introduced the term Pangaea, meaning "all land," to refer to the supercontinent from which all of the current continents separated.

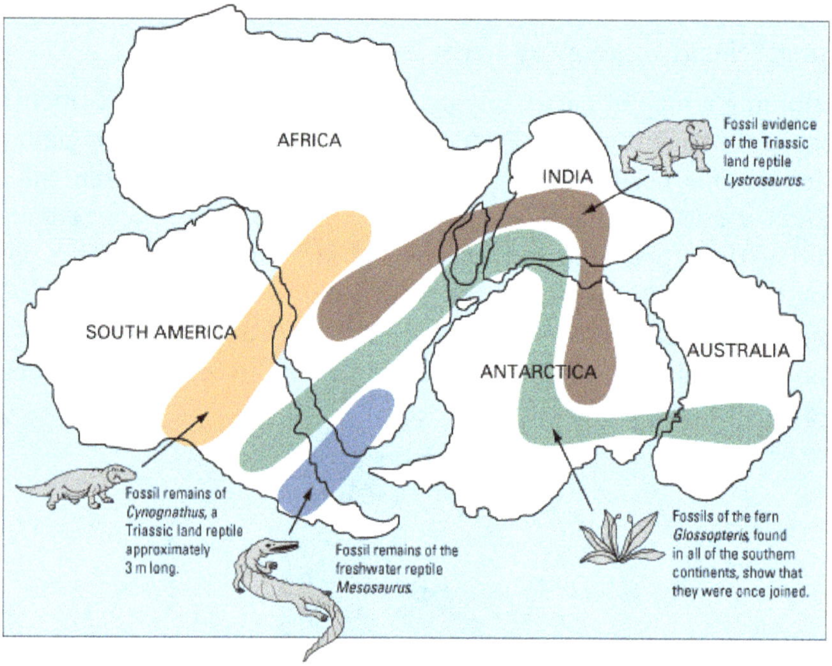

Figure 4.1.2 Distribution of similar fossils across the continents, suggesting they were once connected into a single supercontinent (Steven Earle, "Physical Geology").

Wegener diligently pursued his theory by extensively researching in libraries, seeking advice from others, and conducting observations in order to find compelling evidence to substantiate it. Wegener placed significant emphasis on the alignment of continents and fossil evidence. He also relied on the correlation of geological patterns across oceans, such as the similarity of sedimentary strata in South America and Africa (Fig. 4.1.3), the matching coalfields in North America and Europe, and the resemblance of the mountains in Atlantic Canada and northern Britain in terms of their shape and rock composition.

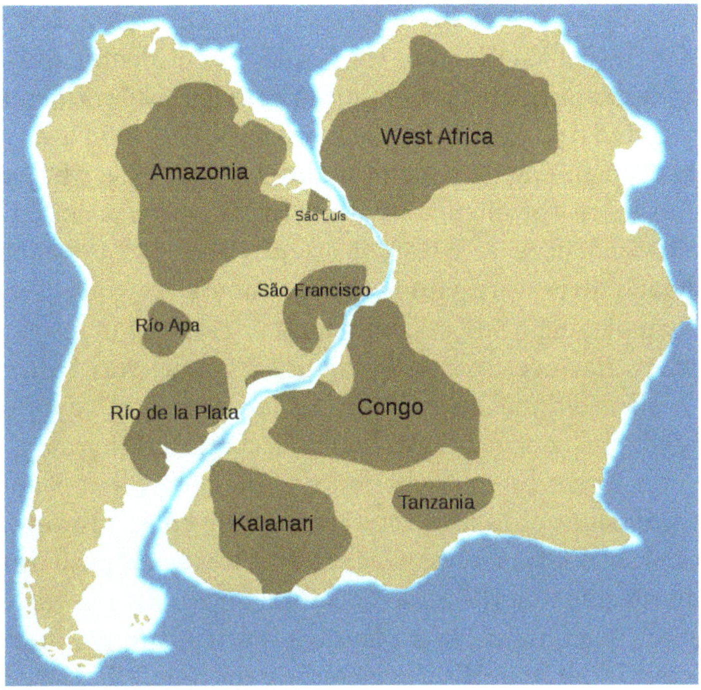

Figure 4.1.3 Matching geological formations spanning from South America to Africa (By Woudloper – Own work, [CC BY-SA 3.0], via Wikimedia Commons).

Wegener used multiple lines of evidence to support his theory of continental drift. Besides the physical fit of the continents and the presence of fossils, he also considered the matching geological patterns across oceans. For example, he observed that sedimentary strata in South America were similar to those in Africa, North American coalfields matched those in Europe, and the mountains in Atlantic Canada resembled those in northern Britain in terms of shape and rock composition.

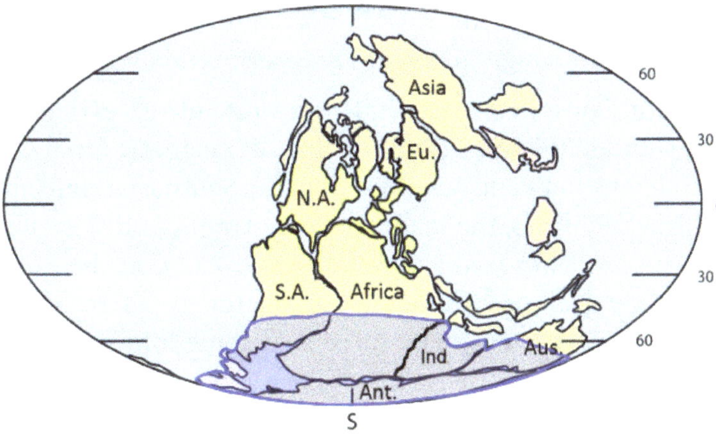

Figure 4.1.4 Extent of glaciation (shaded area) on Pangaea (Modified by PW from Steven Earle, "Physical Geology").

Wegener initially presented his theories in 1912 through a concise publication titled Die Entstehung der Kontinente (The Origin of Continents), and later in 1915 in Die Entstehung der Kontinente und Ozeane (The Origin of Continents and Oceans). He extensively edited this work multiple times until 1929, and it was subsequently translated into French, English, Spanish, and Russian. Nevertheless, despite the abundance of evidence, the alignment of the continents was not flawless, and the geological correspondences were not consistently accurate. It was observed in the 1960s that although there were gaps in the continental fit when using the present-day coastlines, employing a 500 m depth contour resulted in a significantly more precise alignment. However, the most significant issue was Wegener's inability to formulate a plausible mechanism for the movement of continents. Wegener hypothesized that the continents resembled icebergs resting on a denser crust. However, the only forces he could attribute to the movement of continents were poleflucht, the influence of Earth's rotation pushing objects towards the equator, and the lunar and solar tidal forces, which exert a westward push on objects. It was promptly demonstrated that these forces were significantly insufficient to displace continents, and in the absence of a plausible mechanism, Wegener's theory was promptly rejected by the majority of geologists at the time. Alfred Wegener's demise occurred in Greenland in 1930 during his research on glaciation and climate. At the time of his demise, his beliefs were cautiously embraced by just a small faction of geologists, and unequivocally dismissed by the majority. However, in a matter of a few decades, a significant transformation was about to occur.

4.2 PALEOMAGNETIC EVIDENCE FOR PLATE TECTONICS

*Modified from "Physical Geology" by Steven Earle**

Although Alfred Wegener would not live to see it, his theory of plate tectonics would gradually gain acceptance within the scientific community as more evidence began to accumulate. Some of the most important evidence came from the study of **paleomagnetism**, or changes in Earth's magnetic field over millions of years.

Earth's magnetic field is defined by the North and South Poles that align generally with the axis of rotation (Figure 4.2.1). The lines of magnetic force flow into Earth in the Northern Hemisphere and out of Earth in the Southern Hemisphere. Because of the shape of the field lines, the magnetic force trends at different angles to the surface in different locations (red arrows of Figure 4.2.1). At the North and South Poles, the force is vertical. Anywhere on the equator the force is horizontal, and everywhere in between, the magnetic force is at some intermediate angle to the surface.

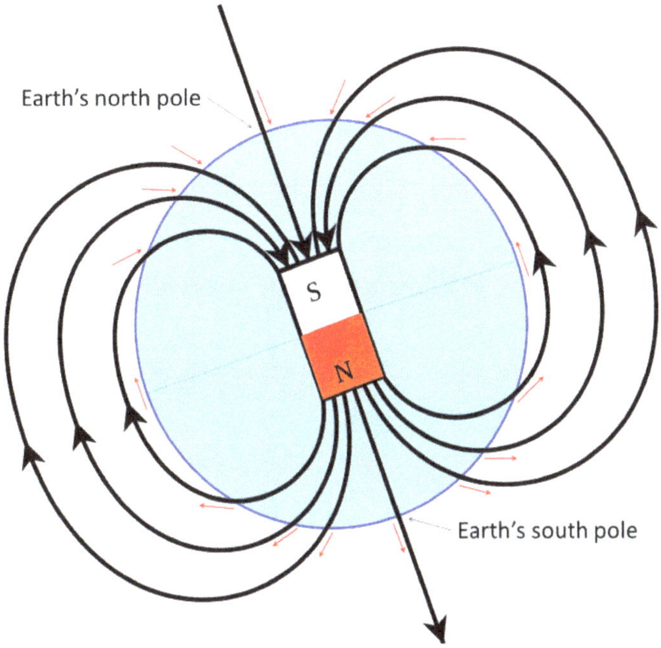

Figure 4.2.1 Depiction of Earth's magnetic field as a bar magnet coinciding with the core. The south pole of such a magnet points to Earth's North Pole. The red arrows represent the orientation of the magnetic field at various locations on Earth's surface (Steven Earle after: http://upload.wikimedia.org/wikipedia/ commons/ 1/17/Earths_Magnetic_Field_ Confusion.svg).

In its fluid form, the minerals that make up magma are free to move in any direction and take on any orientation. But as the magma cools and solidifies, movement ceases and the mineral orientation and position become fixed. As the mineral magnetite (Fe_3O_4) crystallizes from magma, it becomes magnetized with an orientation parallel to that of Earth's magnetic field at that time, similar to the way a compass needle aligns with the magnetic field to point north. This magnetic record in the rock is called **remnant magnetism**. Rocks like basalt, which cool from a high temperature and commonly have relatively high levels of magnetite, are particularly susceptible to being magnetized in this way, but even sediments and sedimentary rocks, as long as they have small amounts of magnetite, will take on remnant magnetism because the magnetite grains gradually become reoriented following deposition. By studying both the horizontal and vertical components of the remnant magnetism, one can tell not only the direction to magnetic north at the time of the rock's formation, but also the latitude where the rock formed relative to magnetic north.

In the early 1950s, a group of geologists from Cambridge University, including Keith Runcorn, Edward Irving and several others, started looking at the remnant magnetism of Phanerozoic British and European volcanic rocks, and collecting paleomagnetic data. They found that rocks of different ages sampled from generally the same area showed quite different apparent magnetic pole positions (green line, Figure 4.2.2). They initially assumed that this meant that Earth's magnetic field had, over time, departed significantly from its present position, which is close to the rotational pole.

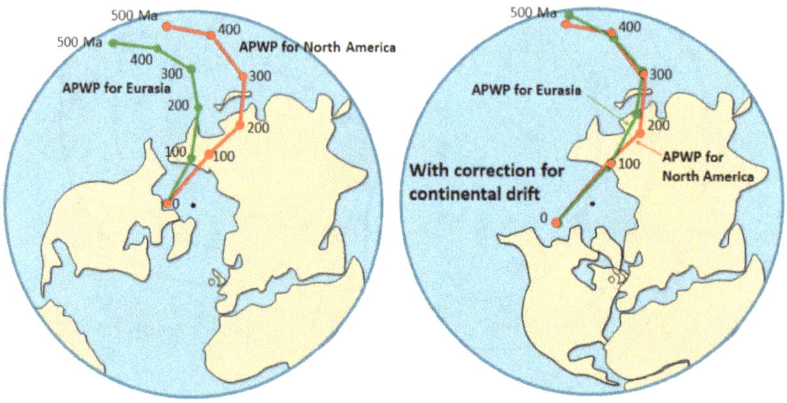

Figure 4.2.2 Polar wandering curves. Curves from Eurasia and North America seem to show that the north magnetic pole was located in two places simultaneously throughout history (left). However, if the continents are rearranged into Pangaea, the two curves overlap, showing that it is the continents than have moved, not the pole (right) (Steven Earle, "Physical Geology").

The curve defined by the paleomagnetic data was called a **polar wandering path** because Runcorn and his colleagues initially thought that their data represented actual movement of the magnetic poles (since geophysical models of the time suggested that the magnetic poles did not need to be aligned with the rotational poles). We now know that the magnetic data define movement of continents, and *not* of the magnetic poles, so we call it an ***apparent*** **polar wandering path** (APWP). Runcorn and colleagues soon extended their work to North America, and this also showed *apparent* polar wandering, but the results were not consistent with those from Europe (Figure 4.2.2). For example, the 200 Ma pole for North America placed somewhere in China, while the 200 Ma pole for Europe placed in the Pacific Ocean. Since there could only have been one pole position at 200 Ma, this evidence strongly supported the idea that North America and Europe had moved relative to each other since 200 Ma. Subsequent paleomagnetic work showed that South America, Africa, India, and Australia also have unique polar wandering curves. Rearranging the continents based on their positions in Pangaea caused these wandering curves to overlap, showing that the continents had moved over time.

Additional evidence for movement of the continents came from analysis of **magnetic dip**. Recall from Figure 4.2.1 that the angle of the magnetic field changes as a function of latitude, with the field directed vertically downwards at the north pole, upwards at the south pole, and horizontal at the equator. Every latitude between the equator and the poles will have a corresponding angle between horizontal and vertical (red arrows, Figure 4.2.1). By looking at the dip angle in rocks, we can determine the latitude at which those rocks were formed. Combining that with the age of the rocks, we can trace the movements of the continents over time. For example, at around 500 Ma, what we now call Europe was south of the equator, and so European rocks formed then would have acquired an upward-pointing magnetic field orientation (Figure 4.2.3). Between then and now, Europe gradually moved north, and the rocks forming at various times acquired steeper and steeper *downward-pointing* magnetic orientations.

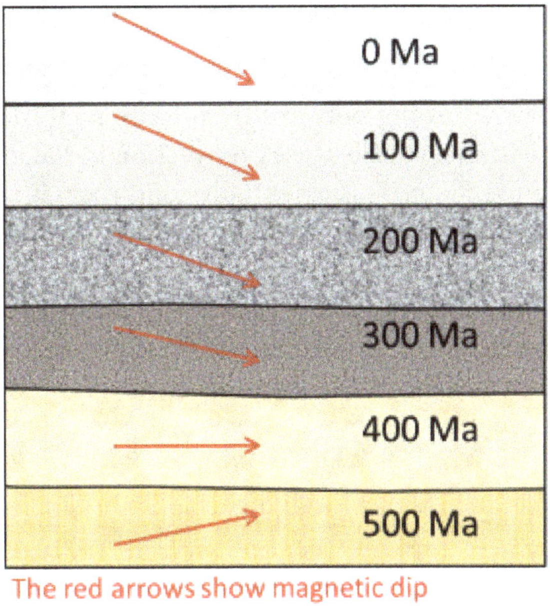

Figure 4.2.3 Hypothetical magnetic dip angles from layers of rock. This rock would have been south of the equator 500 Ma, at the equator 400 Ma, and since then has been moving further north (Steven Earle, "Physical Geology").

This paleomagnetic work of the 1950s was the first new evidence in favor of continental drift, and it led a number of geologists to start thinking that the idea might have some merit.

4.3 MECHANISMS FOR PLATE MOTION

*Modified from "Physical Geology" by Steven Earle**

In section 4.1 we learned that one of the reasons that Wegener's ideas of continental drift were initially rejected by the scientific community was that he could not provide a plausible mechanism for plate motion. However, with all that we have learned about the processes occurring in the Earth's interior since then, there is still some debate about the actual forces that make the plates move. One side in the argument holds that the plates are only moved by the traction caused by mantle convection. The other side holds that traction plays only a minor role and that two other forces, ridge push and slab pull, are more important. Some argue that the real answer lies somewhere in between.

To understand **mantle convection**, imagine a pot of water on a hot stove. The water at the bottom of the pot near the heat source becomes hot and expands, making it lighter (less dense) than the water above. The hot, low density water rises, and cooler, denser water sinks and flows in from the sides. This water then gets heated and rises, and the cycle continues. This creates a circular pattern of rising and sinking water called a **convection cell.** (To test this, try sprinkling a few flakes of spice in the center of a rapidly boiling pot of water. The flakes will move outwards to the edge of the pot as warmer water rises and pushes them aside).

Heat is continuously flowing outward from Earth's interior, and the transfer of heat from the core to the mantle causes convection in the mantle (Figure 4.3.1). Even though the mantle material is essentially solid rock, it is sufficiently plastic (fluid) to slowly flow (at rates of centimeters per year) as long as a steady force is applied to it. This convection is a driving force for the movement of tectonic plates, as the horizontal movements of mantle under the crust drag the plates with them. At places where convection currents in the mantle are moving upward, new lithosphere forms and the plates move apart (diverge). Where two plates are converging (and the convective flow is downward), one plate will be **subducted** (pushed down) into the mantle beneath the other.

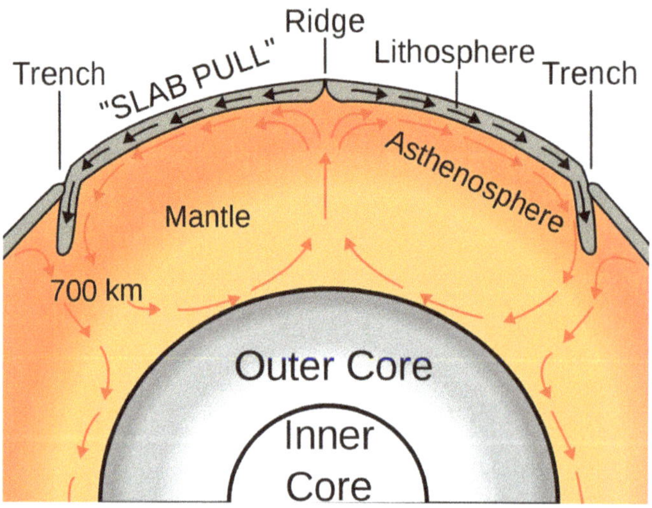

Figure 4.3.1 Convection cells in the mantle (By Surachit [GFDL (http://www.gnu.org/copyleft/fdl.html) or CC BY-SA 3.0], via Wikimedia Commons).

The **ridge push/slab pull** model also relies on mantle convection, but in this case it is not simply the traction from the convection cell that moves the plates.

In this model, plates move through a combination of pull from the weight of the subducting edge of the plates, and through the outward pushing of an ocean ridge where magma is rising and forming new crust (Figure 4.3.2).

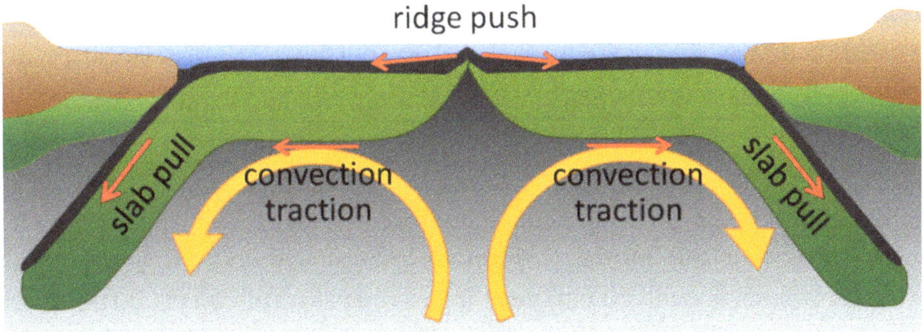

Figure 4.3.2 Models for plate motion mechanisms (Steven Earle, "Physical Geology").

Some compelling arguments in favor of the ridge-push/slab-pull model are as follows: (a) plates that are attached to subducting slabs (e.g., Pacific, Australian, and Nazca Plates) move the fastest, and plates that are not (e.g., North American, South American, Eurasian, and African Plates) move significantly slower; (b) in order for the traction model to apply, the mantle would have to be moving about five times faster than the plates are moving (because the coupling between the partially liquid asthenosphere and the plates is not strong), and such high rates of convection are not supported by geophysical models; and (c) although large plates have potential for much higher convection traction, plate velocity is not related to plate area. Although ridge-push/slab-pull is the favored mechanism for plate motion, it's important not to underestimate the role of mantle convection. Without convection, there would be no ridges to push from because upward convection brings hot buoyant rock to surface. Furthermore, many plates, including our own North American Plate, move along nicely — albeit slowly — without any slab-pull happening.

4.4 PLATES AND PLATE MOTIONS

*Modified from "Physical Geology" by Steven Earle**

The idea of plate tectonics became widely accepted around 1965 as more and more geologists started thinking in these terms. By the end of 1967, Earth's surface had been mapped into a series of plates (Figure 4.4.1). The major plates are Eurasia, Pacific, India, Australia, North America, South America, Africa, and Antarctic. There are also numerous small plates (e.g., Juan de Fuca, Nazca, Scotia, Philippine, Caribbean), and many very small plates or sub-plates. For example the Juan de Fuca Plate is actually three separate plates (Gorda, Juan de Fuca, and Explorer) that all move in the same general direction but at slightly different rates.

The fact that the plates include both crustal material and lithospheric mantle material makes it possible for a single plate to be made up of both oceanic and continental crust. For example, the North American Plate includes most of North America, plus half of the northern Atlantic Ocean. Similarly the South American Plate extends across the western part of the southern Atlantic Ocean, while the European and African plates each include part of the eastern Atlantic Ocean. The Pacific Plate is almost entirely oceanic, but it does include the part of California west of the San Andreas Fault.

Rates of motions of the major plates range from less than 1 cm/year to over 10 cm/year (for comparison, human fingernails grow at around 6 cm/year). The Pacific Plate is the fastest at over 10 cm/year in some areas, followed by the Australian and Nazca Plates. The North American Plate is one of the slowest, averaging around 1 cm/year in the south up to almost 4 cm/year in the north. Plates move as rigid bodies, so it may seem surprising that the North American Plate can be moving at different rates in different places. The explanation is that plates move in a rotational manner. The North American Plate, for example, rotates counter-clockwise; the Eurasian Plate rotates clockwise.

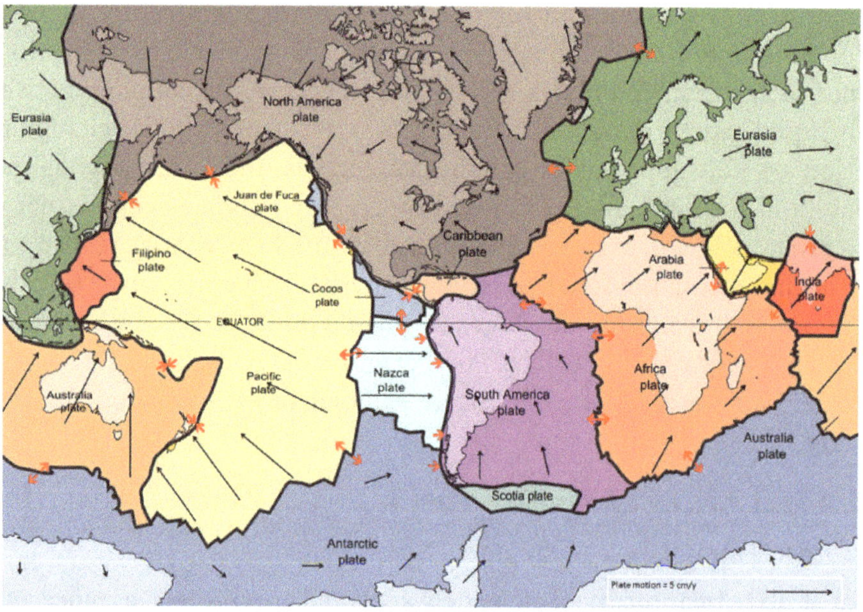

Figure 4.4.1 The major lithospheric plates of Earth. Arrows indicate direction of plate movement, and the length of the arrows represent the speed of plate motion (Steven Earle, "Physical Geology").

As originally described by Wegener in 1915, the present continents were once all part of the supercontinent Pangaea. More recent studies of continental match-ups and the magnetic ages of ocean-floor rocks have enabled us to reconstruct the history of the break-up of Pangaea.

Pangaea began to rift apart along a line between Africa and Asia and between North America and South America at around 200 Ma (Figure 4.4.2). During the same period, the Atlantic Ocean began to open up between northern Africa and North America, and India broke away from Antarctica. At this stage, Pangaea was divided into Laurasia (now Europe, Asia and North America) and Gondwanaland (the southern continents; South America, Africa, India, Australia, and Antarctica). Between 200 and 150 Ma, rifting started between South America and Africa and between North America and Europe, and India separated from Antarctica and moved north toward Asia. By 80 Ma, Africa had separated from South America, and most of Europe had separated from North America. By 50 Ma, Australia had separated from Antarctica, and shortly after that, India collided with Asia.

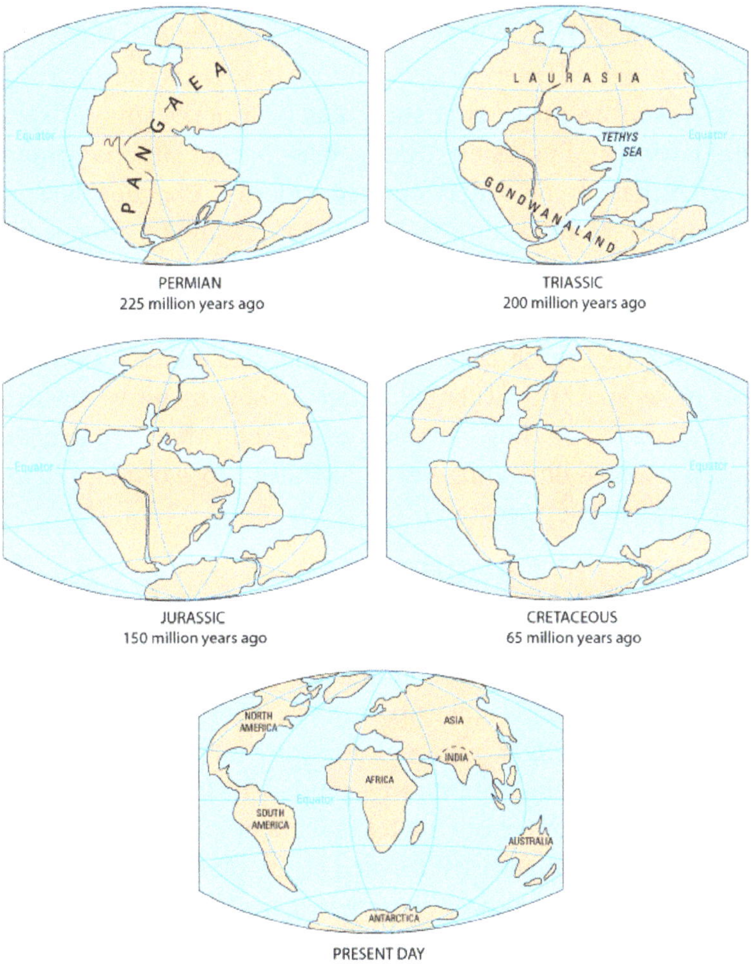

Figure 4.4.2 Movement of the continents over the past 225 million years (USGS, https://pubs.usgs.gov/gip/dynamic/historical.html).

Within the past few million years, rifting has taken place in the Gulf of Aden and the Red Sea, and also within the Gulf of California. Incipient rifting has begun along the Great Rift Valley of eastern Africa, extending from Ethiopia and Djibouti on the Gulf of Aden (Red Sea) all the way south to Malawi.

Over the next 50 million years, it is likely that there will be full development of the east African rift and creation of new ocean floor. Eventually Africa will split apart. There will also be continued northerly movement of Australia and Indonesia. The western part of California (including Los Angeles and part of San Francisco) will split away from the rest of North America, and eventually sail right by the west coast of Vancouver Island, en route to Alaska. Because the oceanic crust formed by spreading on the mid-Atlantic ridge is not currently being subducted (except in the Caribbean), the Atlantic Ocean is slowly getting bigger, and the Pacific Ocean is getting smaller. If this continues without changing for another couple hundred million years, we will be back to where we started, with one supercontinent.

Pangaea, which existed from about 350 to 200 Ma, was not the first supercontinent. In 1966, Tuzo Wilson proposed that there has been a continuous series of cycles of continental rifting and collision; that is, break-up of supercontinents, drifting, collision, and formation of other supercontinents. Pangaea was preceded by Pannotia (600 to 540 Ma), by Rodinia (1,100 to 750 Ma), and by other supercontinents before that.

With all of these plates constantly on the move, they inevitably end up interacting with each other at their plate boundaries. Plates can interact in three ways: they can move apart (divergent boundary), they can move towards each other (convergent boundary), or they can slide past each other (transform boundary). The following sections will examine each of these types of plate boundaries, and the geological features they create.

4.5 DIVERGENT PLATE BOUNDARIES

*Modified from "Physical Geology" by Steven Earle**

Divergent boundaries are spreading boundaries, where new oceanic crust is created to fill in the space as the plates move apart. Most divergent boundaries are located along mid-ocean oceanic ridges (although some are on land). The **mid-ocean ridge** system is a giant undersea mountain range, and is the largest geological feature on Earth; at 65,000 km long and about 1000 km wide, it covers 23% of Earth's surface (Figure 4.5.1). Because the new crust formed at the plate boundary is warmer than the surrounding crust, it has a lower density so it sits higher on the mantle, creating the mountain chain. Running down the middle of the mid-ocean ridge is a **rift valley** 25-50 km wide and 1 km deep. Although oceanic spreading ridges appear to be curved features on Earth's surface, in fact the ridges are composed of a series of straight-line segments, offset at intervals by faults perpendicular to the ridge, called **transform faults**. These transform faults make the mid-ocean ridge system look like a giant zipper on the seafloor (Figure 4.5.2). As we will see in section 4.7, movements along transform faults between two adjacent ridge segments are responsible for many earthquakes.

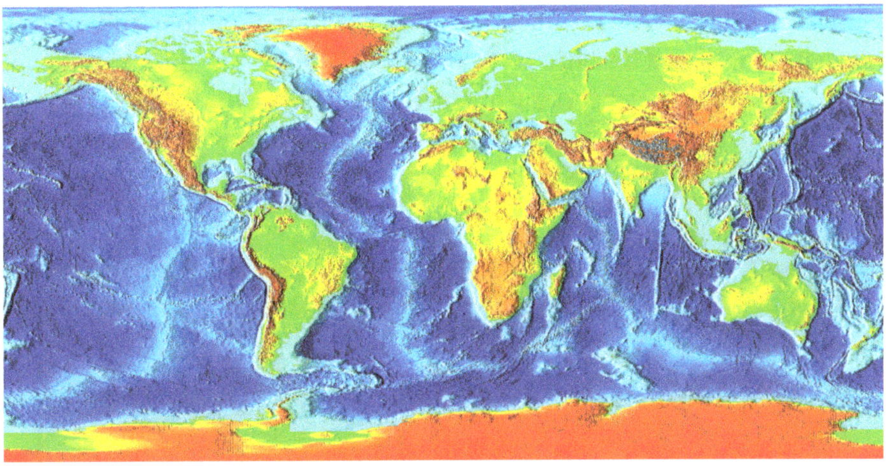

Figure 4.5.1 Ocean floor topography. The mid-ocean ridge system can be seen as the light blue chain of mountains running throughout the oceans (http://www.ngdc.noaa.gov/mgg/image/mggd.gif).

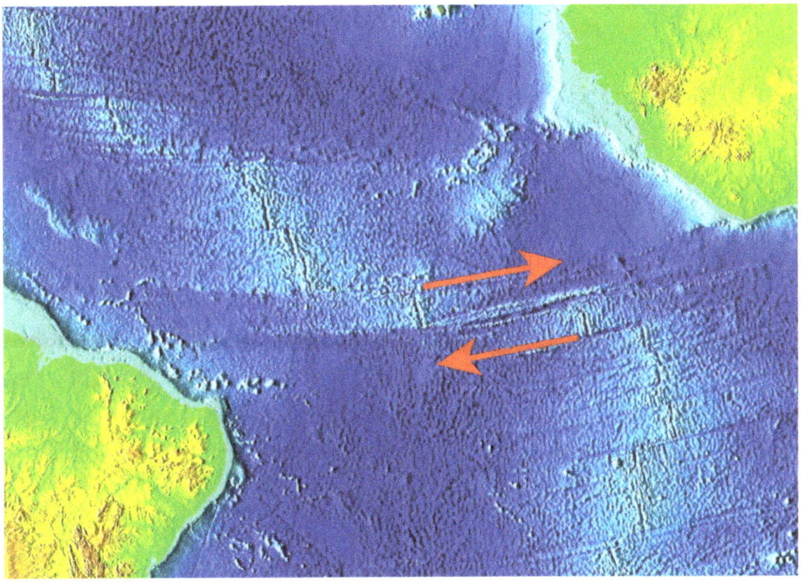

Figure 4.5.2 Closeup of the mid-Atlantic ridge system, showing transform faults perpendicular to the ridge axis. Arrows indicate the direction of plate motion on either side of the fault (USGS, Public domain, via Wikimedia Commons).

The crustal material formed at a spreading boundary is consistently oceanic in nature. This means that it is composed of igneous rock, such as basalt or gabbro, which is rich in ferromagnesian minerals. This rock is created when magma, derived from the partial melting of the mantle due to decompression, moves towards the surface from deeper regions of the mantle (Figure 4.5.3). The partial melting zone adjacent to the ridge crest has a thickness of roughly 60 km and contains magma that is around 10% of the total volume of rock. As a result, the resulting crust is approximately 6 km thick. The magma emerges onto the seafloor and solidifies into pillow basalts, shattered basaltic rock known as breccias, and flowing formations. In certain instances, these formations are found with limestone or chert. Over time, the igneous rock of the oceanic crust accumulates layers of sediment, which then undergoes lithification to form sedimentary rock.

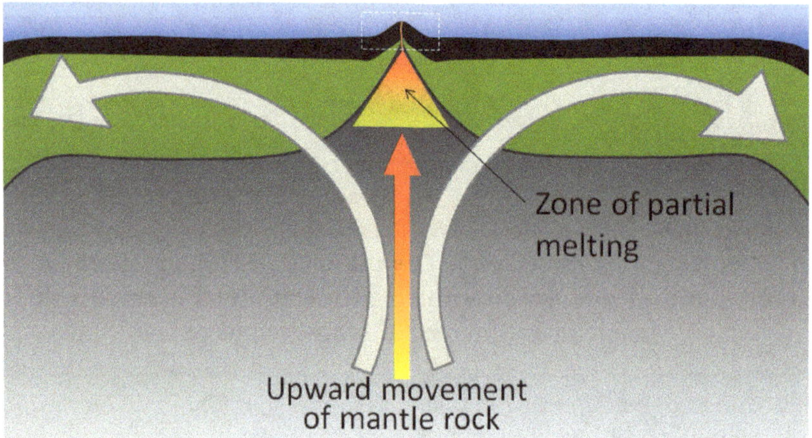

Figure 4.5.3 Mechanism for divergent plate boundaries. The region in the outlined rectangle represent the mid-ocean ridge (Steven Earle, "Physical Geology").

The process of spreading is postulated to initiate inside a continental region through the upward bending or doming of the Earth's crust, which is linked to the presence of a mantle plume or a sequence of mantle plumes beneath. The buoyancy of the material from the mantle plume generates a dome-like structure in the Earth's crust, resulting in its fragmentation. If a huge continent has multiple mantle plumes, the ensuing rifts may align and give rise to the creation of a rift valley, as the Great Rift Valley in eastern Africa. It is hypothesized that this particular type of valley gradually transforms into a linear body of saltwater (like the current Red Sea), and ultimately into a vast expanse of ocean (such as the Atlantic). It is probable that up to 20 mantle plumes, a significant number of which are still present, had a role in the beginning of the separation of Pangaea along the current mid-Atlantic ridge.

Multiple lines of evidence indicate that new oceanic crust is being generated at these seafloor spreading locations.

1. Age of the crust:

By examining the ages of the oceanic crust in close proximity to a mid-ocean ridge, it is evident that the crust is most recent at the spreading center. As one moves away from the diverging boundary in either direction, the crust gradually becomes older. Specifically, the crust ages by around 1 million years for every 20-40 km away from the ridge. Moreover, the distribution of crust age exhibits a very balanced and mirrored pattern on both sides of the ridge (Figure 4.5.4).

The most ancient oceanic crust in the eastern Mediterranean dates back to approximately 280 million years ago, while the oldest sections of the open ocean may be found on both sides of the north Atlantic and are approximately 180 million years old. Despite the fact that certain sections of the continental crust are almost 4,000 million years old, it is remarkable that the oldest seafloor is less than 300 million years old. The reason for this is that any seabed older than that has undergone either subduction (as explained in section 4.6) or has been uplifted to merge with the continental crust. As anticipated, the oceanic crust is relatively new in the vicinity of the spreading ridges (Figure 4.5.4), and there are noticeable variations in the speed of sea-floor spreading across different ridges. The ridges in the Pacific and southern Indian Oceans have broad age ranges, suggesting a high rate of spreading (about 10 cm/year on each side in certain regions), whereas those in the Atlantic and western Indian Oceans experience far slower spreading (less than 2 cm/year on each side in certain regions).

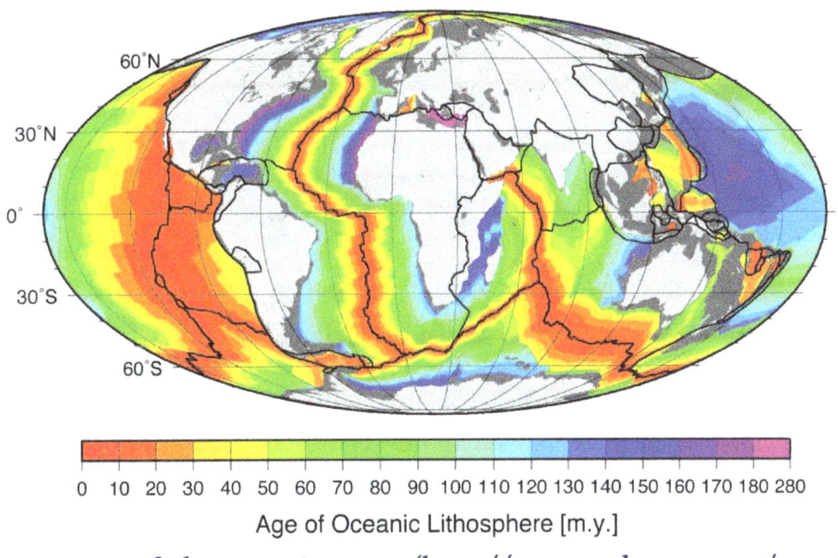

Figure 4.5.4 Age of the oceanic crust (http://www.ngdc.noaa.gov/mgg/ocean_age/data/2008/image/age_oceanic_lith.jpg).

2. Sediment thickness:

Seismic reflection sounding, which is analogous to echo sounding as discussed in section 1.4, allows us to penetrate through seafloor sediments and create maps of the bedrock topography and crustal thickness. Consequently, sediment thicknesses were able to be accurately charted, revealing that whereas sediments near the continents may reach several thousand meters in thickness, they were very thin or absent in the areas of ocean ridges (Figure 4.5.5). This concept is logical when considering the information about the age of the marine crust. The crust becomes older as it moves away from the area where it is formed, resulting in a longer period of time for sediment to accumulate and a thicker layer of sediment. Furthermore, the lower strata of sediment get progressively older as one moves away from the ridge, suggesting that they were originally deposited on the crust during its initial formation at the ridge.

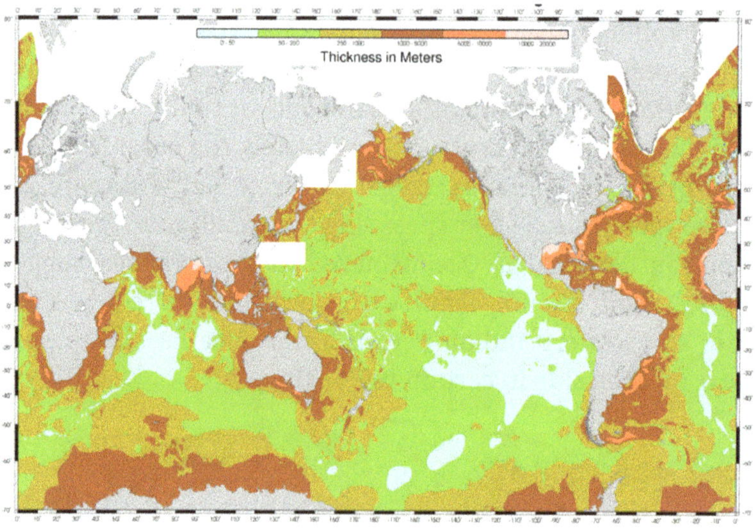

Figure 4.5.5 Seafloor sediment thickness (Modified from https://www.ngdc.noaa.gov/mgg/sedthick/).

3. Heat flow:

Measurements of heat flow rates through the ocean floor indicate that the rates are significantly elevated (about 8 times greater) along the ridges, while they are considerably reduced (about 1/20th of the average) in the trench areas. The regions with high heat flow are associated with the upward movement of hot mantle material during the formation of new crust, while the regions with low heat flow are associated with the downward movement of material at subduction zones.

4. Magnetic reversals:

As discussed in section 4.2, it was observed that rocks have the ability to preserve magnetic data that they obtained throughout their formation. Nevertheless, the Earth's magnetic field undergoes fluctuations over extended periods of geological time. The magnetic field undergoes periodic degradation and subsequent re-establishment, however the exact reasons for this phenomenon remain partially unknown. Upon re-establishment, the object may either retain its original orientation or adopt a reversed polarity. During instances of magnetic reversal, the orientation of a compass would be directed towards the south instead of the north. During the last 250 million years, there have been several hundred instances of magnetic field reversals, and their occurrence has been highly irregular. The geologists have defined the shortest ones to have a duration of only a few thousand years, while the longest one lasted for about 30 million years, specifically during the Cretaceous period (Figure 4.5.6). The current "normal" event has endured for around 780,000 years.

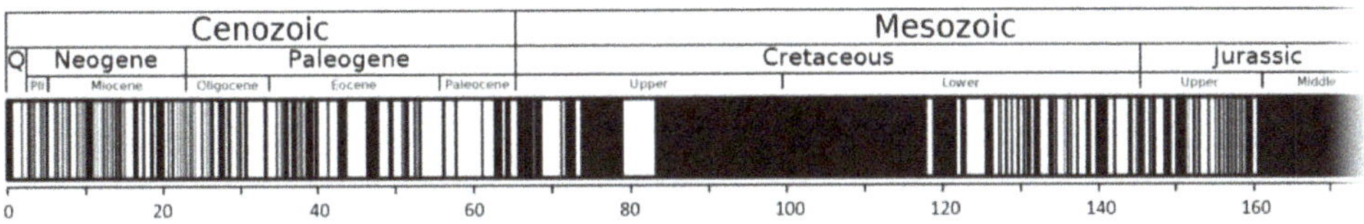

Figure 4.5.6 Magnetic field reversal chronology for the past 170 Ma (Steven Earle after: http://upload.wikimedia.org/wikipedia/en/c/c0/Geomagnetic_polarity_0-169_Ma.svg).

Figure 4.5.7 Pattern of magnetic anomalies in oceanic crust in the Pacific northwest (Steven Earle, "Physical Geology").

From the 1950s forward, scientists began utilizing magnetic readings to investigate the geography of the ocean floor. In 1958, a complete magnetic data set was produced for an offshore area located in British Columbia and Washington State. The scan unveiled an enigmatic arrangement of alternating bands of low and high magnetic intensity in the rocks on the sea floor (Figure 4.5.7). Following research conducted in other parts of the ocean, similar magnetic irregularities were also detected. Of particular significance is the observation that these magnetic patterns exhibit symmetry in relation to ocean ridges. In the 1960s, the Vine-Matthews-Morley (VMM) hypothesis suggested that the patterns found in ridges were connected to magnetic reversals. According to this hypothesis, oceanic crust formed from cooling basalt during a normal event would have the same polarity as the current magnetic field, resulting in a positive anomaly (a black stripe) on the magnetic map of the sea-floor. Conversely, oceanic crust formed during a reversed event would have opposite polarity to the present field, leading to a negative magnetic anomaly (a white stripe). The widths of the anomalies differed based on the spreading rates that are characteristic of the various ridges. The method is depicted in Figure 4.5.8. A fresh crust is generated (panel a) and adopts the prevailing magnetic polarity. As the plates gradually move apart, the magnetic polarity undergoes a reversal. The newly produced crust at the ridge then acquires the opposite polarity, which is represented by the white stripes in Figure 4.5.8. In panel b, the poles have returned to their original state, indicating that the new crust is once again exhibiting normal polarity before drifting away from the ridge. Over time, this process results in the formation of a sequence of parallel, alternating bands of reversals that are symmetrical around the spreading center (panel c).

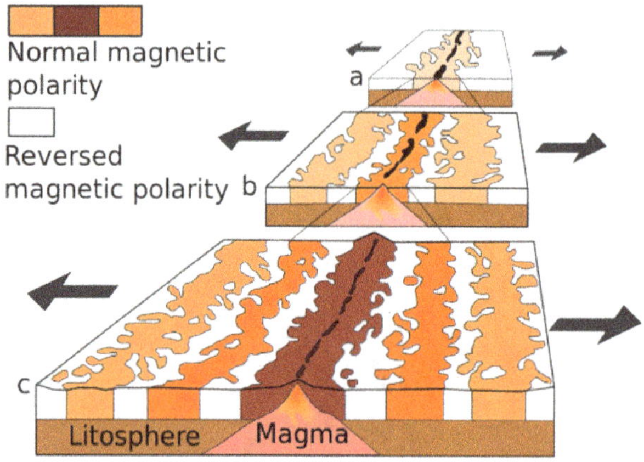

Figure 4.5.8 Formation of alternating patterns of magnetic polarity along a mid-ocean ridge (Steven Earle, "Physical Geology").

4.6 CONVERGENT PLATE BOUNDARIES

*Modified from "Physical Geology" by Steven Earle**

Convergent borders, characterized by the movement of two plates towards one other, can be classified into three types based on the type of crust found on each side of the boundary: oceanic or continental. The three forms of plate boundaries are ocean-ocean, ocean-continent, and continent-continent.

At an ocean-ocean convergent boundary, one of the plates, which consists of oceanic crust and lithospheric mantle, is forcefully pushed or subducted beneath the other plate (as shown in Figure 4.6.1). Frequently, the older and colder plate is denser and undergoes subduction beneath the younger and warmer plate. Typically, there is an oceanic trench formed at the boundary where the crust bends downward. The lithosphere that is being subducted descends into the mantle at a shallow angle near the subduction zone, but at steeper angles further down (up to around 45°). The substantial amount of water contained in the subducting material is expelled when the subducting crust is heated. When it combines with the mantle above it, the crust's melting point is reduced and magma is formed by the process of flux melting, which occurs when water is added to the hot mantle. Magma, being less dense than the mantle material it surrounds, ascends through the mantle and the overlaying oceanic crust to the ocean bottom, giving rise to a series of volcanic islands called an island arc. A mature island arc transforms into a series of comparatively sizable islands, like Japan or Indonesia, due to the continuous extrusion of volcanic material and the accumulation of sedimentary materials around the islands. Earthquakes typically occur at significant depths beneath the seafloor, where the crust that is being forced beneath another crust moves in opposition to the crust above it.

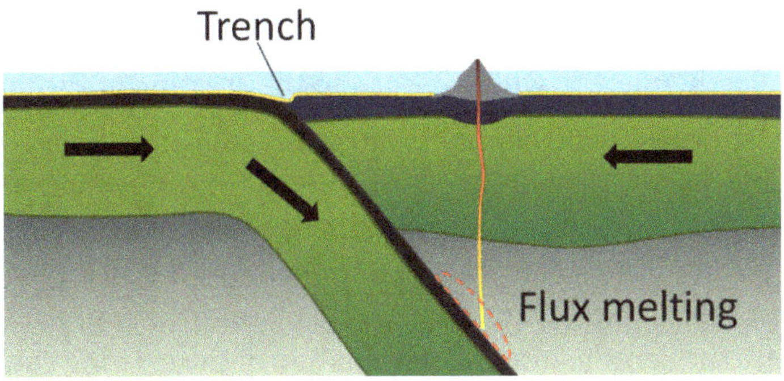

Figure 4.6.1 A trench and volcanic island formed from an ocean-ocean convergent zone (Steven Earle, "Physical Geology").

Two examples of ocean-ocean convergent zones include the subduction of the Pacific Plate south of Alaska, which leads to the formation of the Aleutian Islands, and the subduction of the Pacific Plate under the Philippine Plate, resulting in the creation of the Marianas Trench, which is the deepest portion of the ocean.

At an ocean-continent convergent boundary, the more compact oceanic plate is subducted beneath the less compact continental plate, similar to what occurs at an ocean-ocean boundary. The sediment deposited on the bottom is pushed upwards into a wedge-shaped formation called an accretionary wedge. The compression forces within the continental plate cause thrusting to occur (see Figure 4.6.2). The magma generated near the subduction zone ascends towards the lower part of the continental crust, causing the crustal rock to undergo partial melting. The magma that is formed moves upwards through the Earth's crust, creating a series of mountains that have several volcanoes. Similar to an ocean-ocean boundary, the process of subducting crust can result in the formation of a trench that extends deeply and runs parallel to the coastline.

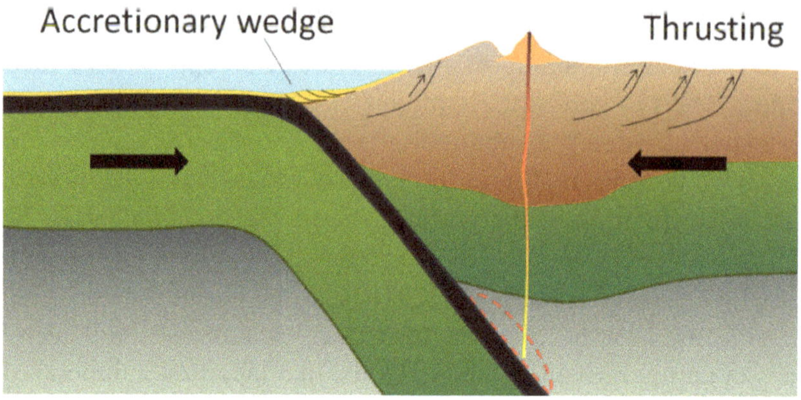

Figure 4.6.2 A trench and volcanic mountains formed from an ocean-continent convergent zone (Steven Earle, "Physical Geology").

Two examples of ocean-continent convergent boundaries include the subduction of the Nazca Plate beneath South America, resulting in the formation of the Andes Mountains and the Peru Trench, and the subduction of the Juan de Fuca Plate beneath North America, which has led to the creation of the Cascade Range.

A continent-continent collision refers to the event where a continent or a huge island, which has been carried along with subducting oceanic crust, comes into contact with another continent (Figure 4.6.3). The collision between the continental materials will not result in subduction due to their low density, primarily made of light continental rocks. However, the oceanic plate's root will eventually detach and descend into the mantle. Intense deformation of the existing continental rocks results in their upward movement, leading to the development of mountains.

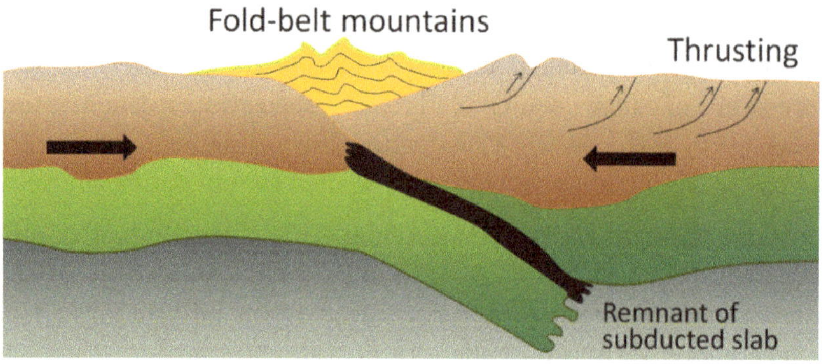

Figure 4.6.3 Mountains formed from a continent-continent convergent zone (Steven Earle, "Physical Geology").

Instances of continent-continent convergent boundaries include the collision between the India Plate and the Eurasian Plate, resulting in the formation of the Himalaya Mountains, as well as the collision between the African Plate and the Eurasian Plate, leading to the creation of a sequence of mountain ranges stretching from the Alps in Europe to the Zagros Mountains in Iran.

4.7 TRANSFORM PLATE BOUNDARIES

*Modified from "Physical Geology" by Steven Earle**

Transform boundaries exist where one plate slides past another without production or destruction of crustal material. As explained in section 4.5, most transform faults connect segments of mid-ocean ridges and are thus ocean-ocean plate boundaries. Some transform faults connect continental parts of plates. An example is the San Andreas Fault, which connects the southern end of the Juan de Fuca Ridge with the northern end of the East Pacific Rise (ridge) in the Gulf of California (Figure 4.7.1). The part of California west of the San Andreas Fault and all of Baja California are on the Pacific Plate. Transform faults do not just connect divergent boundaries. For example, the Queen Charlotte Fault connects the north end of the Juan de Fuca Ridge, starting at the north end of Vancouver Island, to the Aleutian subduction zone.

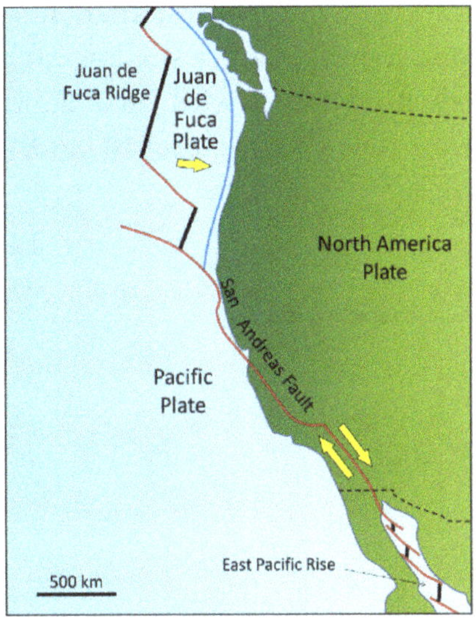

Figure 4.7.1 Transform faults along the U.S. west coast (Steven Earle, "Physical Geology").

As we will see in the next section, earthquakes are common along transform faults, as the two plates slide past each other.

4.8 EARTHQUAKES AND PLATE TECTONICS

*Modified from "Physical Geology" by Steven Earle**

An earthquake refers to the seismic activity resulting from the rupture and subsequent shift of rocks beneath the Earth's surface.

When a mass of rock is subjected to stress, it undergoes deformation. Once the rock reaches its limit of deformation, it fractures and the two sides move horizontally in opposite directions. Due to its inherent strength, rock is capable of enduring substantial deformation without fracturing, unlike loose sand. However, it is important to note that every rock possesses a specific deformation threshold, and it will fracture (break) once that threshold is surpassed. At that juncture, when it comes to rocks located within the Earth's crust, the rock fractures and experiences displacement along the rupture surface. The magnitude of an earthquake is determined by the size of the area that breaks, known as the rupture surface, and the average amount of sliding or displacement that occurs.

The majority of earthquakes occur in close proximity to plate boundaries, however they may not occur directly on a boundary and may not even be associated with a pre-existing fault. Figure 4.8.1 displays the global distribution of earthquakes. The correlations between earthquakes and plate borders are readily observable. Earthquakes are frequent along divergent boundaries such as the mid-Atlantic ridge and the East Pacific Rise. However, they are limited to a confined area around the ridge and generally occur at depths of less than 30 km. Transform faults, like the San Andreas Fault, frequently experience shallow earthquakes. Earthquakes are quite prevalent along subduction zones, with their depth increasing significantly on the side of the subduction zone that is closer to land.

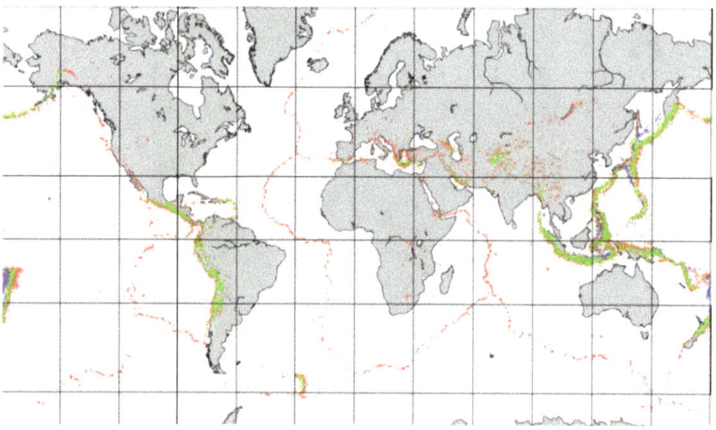

Figure 4.8.1 Global distribution of earthquakes. Red dots indicate shallow earthquakes (<33 km deep), green and blue indicate deep earthquakes (Steven Earle, "Physical Geology").

Earthquakes occur frequently in a few specific areas inside tectonic plates. Some earthquakes are associated with the accumulation of tension caused by continental rifting or the transfer of stress from other areas, while others have unclear causes. Notable intraplate earthquake locations include the Great Rift Valley in Africa, the Tibet region in China, and the Lake Baikal area in Russia.

Earthquakes occurring at divergent and transform boundaries

Figure 4.8.2 provides a closer look at magnitude (M) 4 and larger earthquakes in an area of divergent boundaries in the mid-Atlantic region near the equator. Here, as we saw in section 4.5, the segments of the mid-Atlantic ridge are offset by some long transform faults. Most of the earthquakes are located along the transform faults, rather than along the spreading segments, although there are clusters of earthquakes at some of the ridge-transform boundaries. Some earthquakes do occur on spreading ridges, but they tend to be small and infrequent because of the relatively high rock temperatures in the areas where spreading is taking place. Earthquakes along divergent and transform boundaries tend to be shallow, as the crust is not very thick.

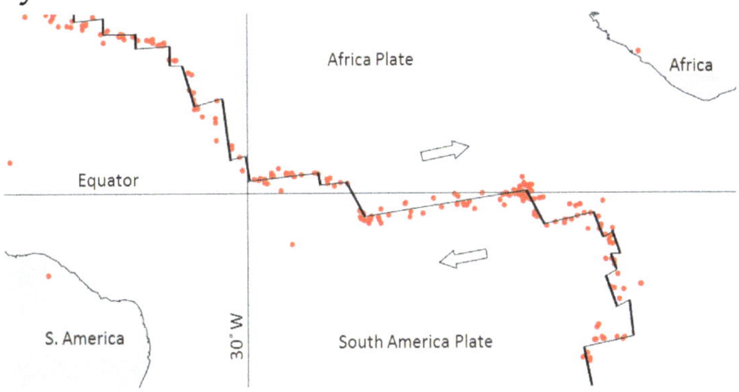

Figure 4.8.2 Earthquake activity along the mid Atlantic ridge (Steven Earle, "Physical Geology").

Earthquakes at Convergent Boundaries

The distribution and depths of earthquakes in the North Pacific are shown in Figure 4.8.3. In this region, the Pacific Plate is subducting beneath the North America Plate, creating the Aleutian Trench and the Aleutian Islands. Shallow earthquakes are common along the trench, but there is also significant earthquake activity extending down several hundred kilometers, as the subducting plate continues to interact at depth with the overriding plate. The earthquakes get deeper with distance from the trench; note in the left panel in Figure 4.8.3 that as you move along the transect from point a to point b, there is a trend of increasing earthquake depth. This reveals that it is the Pacific Plate that is moving northwards and being subducted.

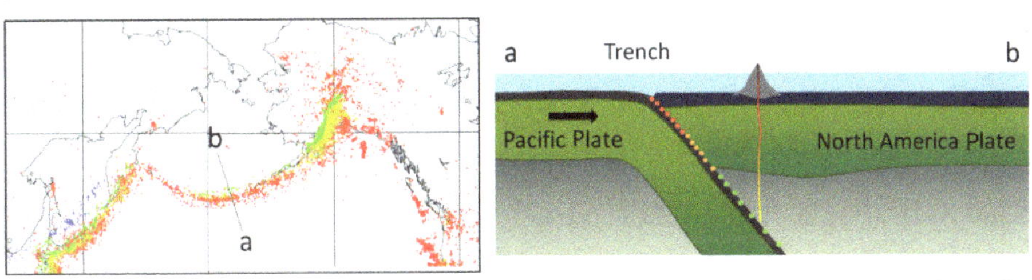

Figure 4.8.3 Earthquake activity along a convergent boundary at the Aleutian Islands. Red dots indicate shallow earthquakes, green and blue indicate deeper earthquakes (Steven Earle, "Physical Geology").

The distribution of earthquakes in the area of the India-Eurasia plate boundary is shown in Figure 4.8.4. This is a continent-continent convergent boundary, and it is generally assumed that although the India Plate continues to move north toward the Asia Plate, there is no actual subduction taking place. There are transform faults on either side of the India Plate in this area.

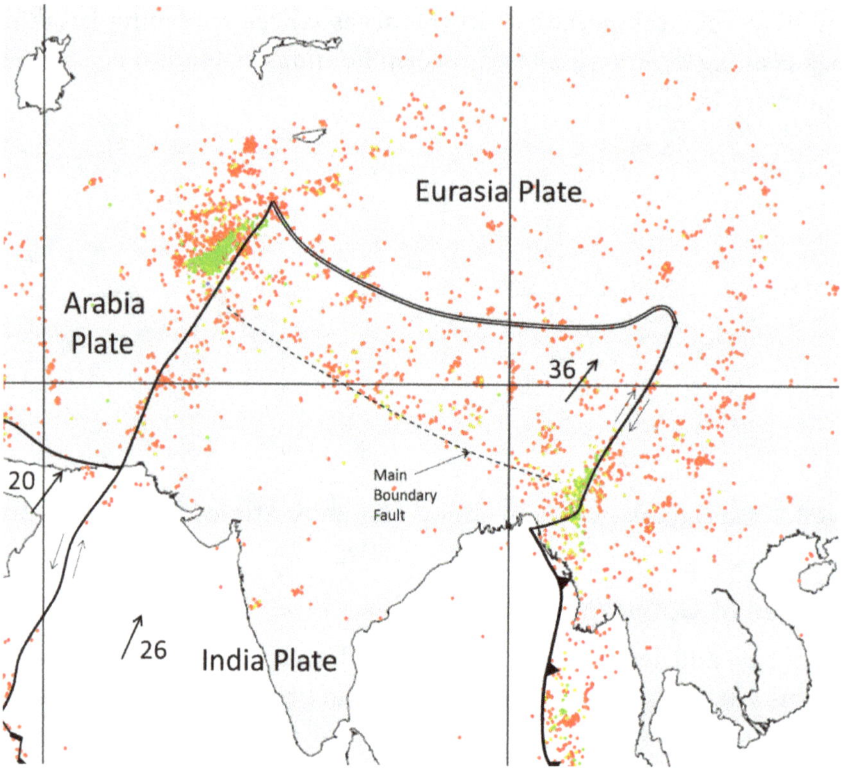

Figure 4.8.4 The distribution of earthquakes in the area of the India-Eurasia plate boundary (Steven Earle, "Physical Geology").

The entire northern India and southern Asia region is very seismically active. Earthquakes are common in northern India, Nepal, Bhutan, Bangladesh and adjacent parts of China, and throughout Pakistan and Afghanistan. Many of the earthquakes are related to the transform faults on either side of the India Plate, and most of the others are related to the significant tectonic squeezing caused by the continued convergence of the India and Asia Plates. That squeezing has caused the Asia Plate to be thrust over top of the India Plate, building the Himalayas and the Tibet Plateau to enormous heights.

4.9 SEAMOUNTS AND HOT SPOTS

*Modified from "Physical Geology" by Steven Earle**

The ocean floor is dotted with **seamounts**, some isolated and some in chains. Seamounts are underwater volcanoes, and most are much younger than the oceanic crust on which they formed. If a seamount gets large enough to break the ocean surface, it becomes a volcanic island. Some seamounts are formed from magma rising at a divergent boundary, and as the plates move apart, the seamounts move with them, which can result in a seamount chain. Other seamounts form from the rising magma at an ocean-ocean subduction zone; these include the Aleutians, extending from Alaska to Russia, and the Lesser Antilles in the eastern part of the Caribbean. Sometimes the crust on which an island or seamount sits will subside, taking the seamount with it. As this happens, the top of the seamount can become eroded flat, and these flat-topped seamounts are then called **tablemounts** or **guyots**.

However, some seamounts are formed far away from plate boundaries, in places where we would not usually expect much volcanic activity. Some seamounts and ocean islands are formed above a **mantle plume** or **hot spot** – a place where hot mantle material rises in a stationary and semi-permanent plume, and affects the overlying crust. Mantle plumes are thought to rise at approximately 10 times the rate of mantle convection. The ascending column may be on the order of kilometers to tens of kilometers across, but near the surface it spreads out to create a mushroom-style head that is several tens to over 100 kilometers across. Near the base of the lithosphere (the rigid part of the mantle), the mantle plume (and possibly some of the surrounding mantle material) partially melts to form magma that rises to feed volcanoes.

A great example of seamounts created from a hot spot includes the Hawaiian and Emperor Seamount island chains in the Pacific Ocean (Figure 4.9.1). The oldest of the Hawaiian/Emperor seamounts is dated at around 80 Ma, and it is situated on oceanic crust aged around 90 to 100 Ma. The volcanic rock making up these islands gets progressively younger toward the southeast, culminating with the island of Hawaii itself, which consists of rock that is almost all younger than 1 Ma. It appears that a stationary plume of hot upwelling mantle material is the source of the Hawaiian volcanism, and that the ocean crust of the Pacific Plate is moving toward the northwest over this hot spot. A seamount will be formed through volcanic activity over the hot spot, then the plate will move and displace the seamount before the hot spot produces the next seamount, and so on. In this way, over time, the seamounts are formed in chains. Near the Midway Islands, the chain takes a pronounced change in direction, from northwest-southeast for the Hawaiian Islands to nearly north-south for the Emperor Seamounts. This change is widely ascribed to a change in direction of the Pacific Plate moving over the stationary mantle plume, but it is also possible that the Hawaiian mantle plume has not actually been stationary throughout its history, and in fact moved at least 2,000 km south over the period between 81 and 45 Ma.

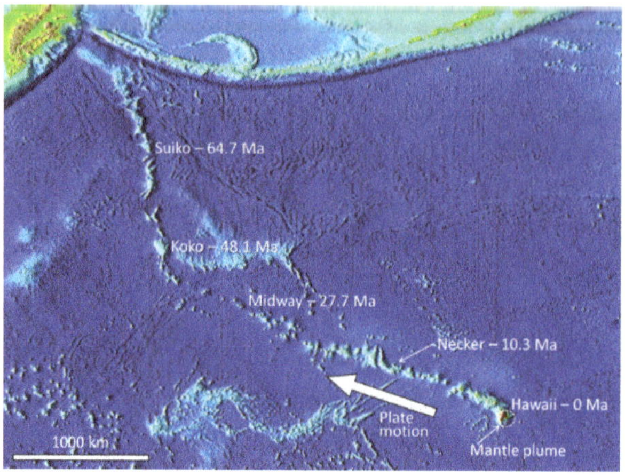

Figure 4.9.1 The Hawaiian Islands/Emperor Seamount chain, with ages of selected structures. This chain has formed as the Pacific Plate moved northwest over a hot spot (Steven Earle, "Physical Geology").

Since most mantle plumes are beneath the oceans, the early stages of volcanism typically take place on the seafloor. Over time, very large islands may form like those in Hawaii. In fact, if you measure it from its base on the seafloor to its summit, Mauna Loa on the island of Hawaii is the largest mountain on Earth, rising 9700 m (in comparison, the elevation of the summit of Mt. Everest is 8848 m). While the island of Hawaii is the youngest in the chain, there is actually a new volcano named Loihi, that is still submerged at a depth of 980 m SE of Hawaii, and may one day become a new Hawaiian island when it emerges 10,000 – 100,000 years from now.

There is evidence of many such mantle plumes around the world. Most are within the ocean basins, including places like Hawaii, Iceland, and the Galapagos Islands, but some are under continents. One example is the Yellowstone hot spot in the west-central United States, and another is the one responsible for the Anahim Volcanic Belt in central British Columbia. It is evident that mantle plumes are very long-lived phenomena, lasting for at least tens of millions of years, possibly for hundreds of millions of years in some cases.

4.10 CORAL REEFS

*Modified from "Physical Geology" by Steven Earle**

Although it may appear unusual to be addressing coral reefs in a geology section, the presence of calcareous skeletons produced by various coral species makes them equally captivating from a geological perspective as they are from a biological standpoint. Corals thrive in optimal conditions characterized by warm, transparent, tropical water, which is sufficiently shallow to allow for photosynthesis by the algae residing within the coral tissues. Due of the necessity of light, fresh coral frequently develops on the surface of the calcified remains of older corals.

During the 1830s, Charles Darwin conducted investigations on several coral reefs and proposed a hypothesis suggesting that they demonstrate a sequential development from one form to another. The researcher analyzed fringing reefs, barrier reefs, and atolls, which are commonly seen near oceanic islands (Figure 4.10.1). Fringing reefs are coral reefs that are in close proximity to or directly related to the shoreline. Barrier reefs are submerged reefs located offshore and are distinguished from the land by a wide stretch of water, typically a lagoon. Atolls are ring-shaped coral reefs that encircle a lagoon, lacking a central landmass within the lagoon. Darwin hypothesized that reefs evolved from fringing, to barrier, to atolls as the land mass gradually sank. Nevertheless, he was unable to provide a rationale for the submergence of volcanic islands. Today, it is widely accepted that Darwin's theory was accurate, and we now understand that islands can sink due to the subsidence of oceanic crust as it moves away from a spreading core, or due to the rising sea level caused by melting glaciers.

Figure 4.10.1 A fringing reef (left), barrier reef (center), and atoll (right) (Public domain, via Wikimedia Commons).

The sequence commences with the formation of a fringing reef adjacent to the coast of an island (Figure 4.10.2). If there is no change in sea level or sinking of the land, numerous reefs will not advance beyond this stage. However, in the event of land subsidence, the corals would gradually descend to a depth where light cannot penetrate (see to section 6.5), resulting in their demise. As the depth of the reef increases, the corals persistently grow vertically, at an approximate rate of 3-5 meters per 1000 years. Over time, a lagoon forms between the reef and the island, transforming the reef into a barrier reef. If the land continues to sink until it is fully inundated, only a circular formation of coral, known as an atoll, will remain, since it has been steadily developing upwards around the central lagoon.

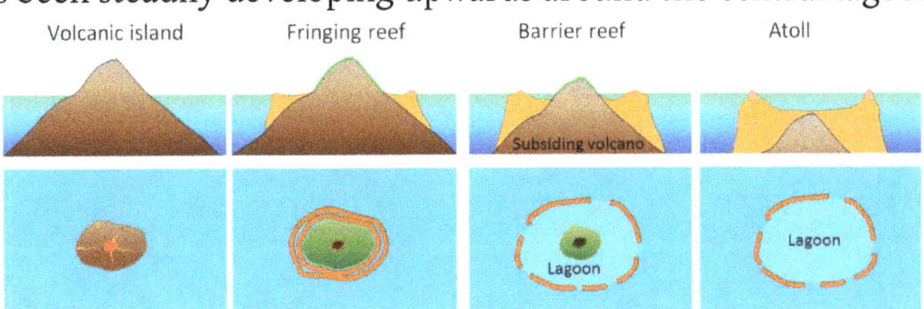

Figure 4.10.2 Steps in the development of coral reefs (Steven Earle, "Physical Geology").

5 Chapter

CHEMICAL OCEANOGRAPHY: CHEMICAL COMPOSITION AND PROCESSES IN THE OCEAN

5.1 PROPERTIES OF WATER

The primary characteristic of the oceans is their composition of water. Water, although seemingly unremarkable due to its abundance, possesses numerous distinctive characteristics that significantly influence worldwide oceanographic and climatological phenomena. The occurrence of these processes is mostly attributed to the formation of hydrogen bonds between water molecules.

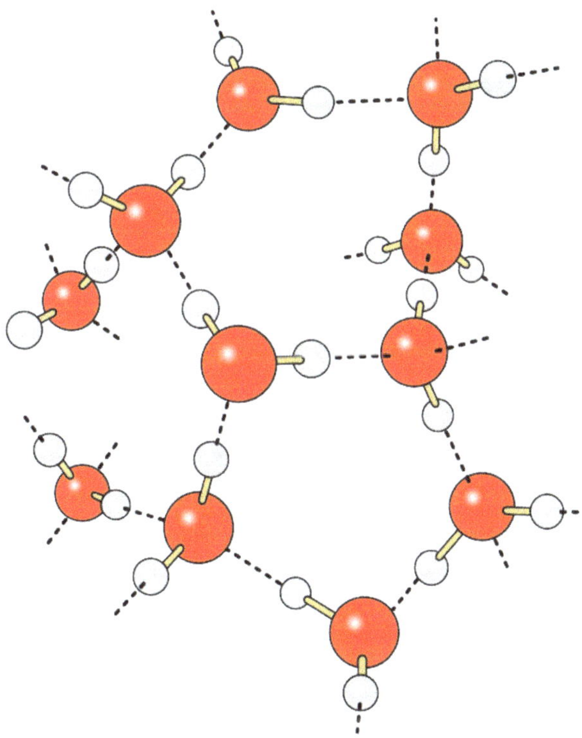

Figure 5.1.1 Hydrogen bonds (dashed lines) between water molecules. Oxygen atoms are shown in red, hydrogen atoms in white (Public domain, via Wikimedia Commons).

A water molecule is composed of two hydrogen atoms and one oxygen atom. Electrons involved in the bonding between atoms are unevenly distributed inside the molecule, resulting in a slight positive charge at the hydrogen ends of water molecules and a slight negative charge at the oxygen end. This uneven distribution of charges makes water a polar molecule. The oxygen atom with a negative charge in the molecule is attracted to the hydrogen atom with a positive charge in an adjacent molecule. The force of attraction, which is rather weak, is referred to as a hydrogen bond (Figure 5.1.1). Without the presence of hydrogen bonding, the temperature at which water would transition into vapor would be -680 C. thus, the existence of liquid water, and thus life, would be impossible on Earth. The hydrogen bonding in water are accountable for several distinctive features of this substance:

1. Water is the only substance to naturally exist in a solid, liquid, and gaseous form under the normal range of temperatures and pressures found on Earth. This is due to water's relatively high freezing and vaporizing points (see below).
2. Water has a **high heat capacity**, which is the amount of heat that must be added to raise its temperature. Specific heat is the heat required to raise the temperature of 1 g of a substance by 1° C. Water has the highest specific heat of any liquid except ammonia (Table 5.1.1).

Table 5.1.1 Specific heat values for a number of common substances

	Specific Heat (calories/g/C°)
Ammonia	1.13
Water	1.00
Acetone	0.51
Grain Alcohol	0.23
Aluminum	0.22
Copper	0.09
Silver	0.06

Water is also one of the most challenging liquids to heat or cool due to its high heat capacity, allowing it to absorb significant amounts of heat without experiencing a substantial increase in temperature. Keep in mind that temperature is a measure of the average energy of motion of the molecules in a substance. The greater the intensity of the motion, the higher the temperature. Water molecules are interconnected through hydrogen bonds, which must be surmounted in order for the molecules to achieve unrestricted movement. When thermal energy is applied to water, it is initially used to disrupt the hydrogen bonds before the temperature of the water can increase. Consequently, a significant portion of the additional heat is absorbed through the disruption of hydrogen bonds, rather than by raising the temperature, resulting in water having a high heat capacity.

Hydrogen bonds confer upon water a high latent heat capacity, which is the amount of heat energy needed for a substance to transition from a solid to a liquid state, or from a liquid to a gaseous state. The latent heat of fusion refers to the amount of heat energy needed to convert a substance from a solid state to a liquid state. In the specific situation of ice melting into water, this requires 80 calories per gram. Ice is in a solid state due to the presence of hydrogen bonding that keep the water molecules arranged in a solid crystal lattice structure. As heat is applied to ice, its temperature increases until it reaches 0 degrees Celsius. At this juncture, any further heat is allocated towards the process of melting the ice by disrupting the hydrogen bonds, rather than towards elevating the temperature. As long as there remains ice, the temperature of the water will remain constant and not rise. The reason why your drink stays cold is because any heat that is absorbed is used to melt the ice, rather than warming up the liquid.

Once all the ice has melted, any extra heat will cause the temperature of the water to rise by 1 degree Celsius for every calorie of heat provided, until it reaches 100 degrees Celsius. At that juncture, any surplus heat is utilized to surmount the hydrogen bonds and transform the liquid water into water vapor, instead of elevating the water temperature. The amount of heat needed to convert liquid water into water vapor is known as the latent heat of vaporization, with a specific value of 540 calories per gram (Figure 5.1.2).

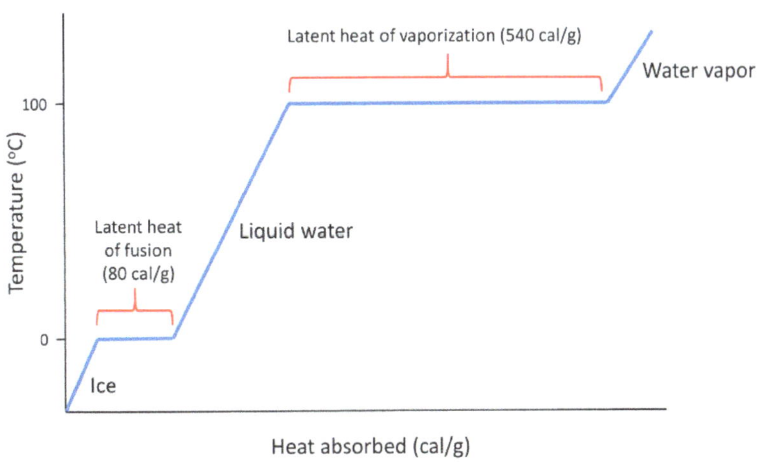

Figure 5.1.2 Latent heat required for phase changes in water. Latent heat of fusion is the heat required to melt ice (80 cal/g), and latent heat of vaporization is the heat required to turn liquid water into water vapor (540 cal/g) (PW).

The substantial heat capacity of water plays a crucial role in maintaining global climatic stability. This is achieved by the gradual absorption and release of heat by the oceans, which effectively prevents abrupt fluctuations in temperature. For further details, please refer to section 8.1. Additionally, this implies that aquatic creatures are not as susceptible to the same abrupt fluctuations in temperature as terrestrial organisms. An organism living in the deep ocean may undergo a temperature variation of no more than 0.5 degrees Celsius throughout its entire lifespan, whereas a species inhabiting land may face temperature fluctuations exceeding 20 degrees Celsius within a single day!

Water has the ability to dissolve a greater number of chemicals compared to any other liquid. This characteristic makes it a "universal solvent", which explains why a large number of substances are dissolved in the ocean. Water exhibits a high affinity for dissolving ionic salts, which are composed of oppositely charged ions like NaCl (Na^+ and Cl^-). Within a water medium, the electrically charged ions exert an attractive force on the polar water molecules. The ions are enveloped by a layer of water molecules, causing the link between the ions to be weakened by a factor of up to 80. When the links between ions are loosened, the substance dissolves (Figure 5.1.3).

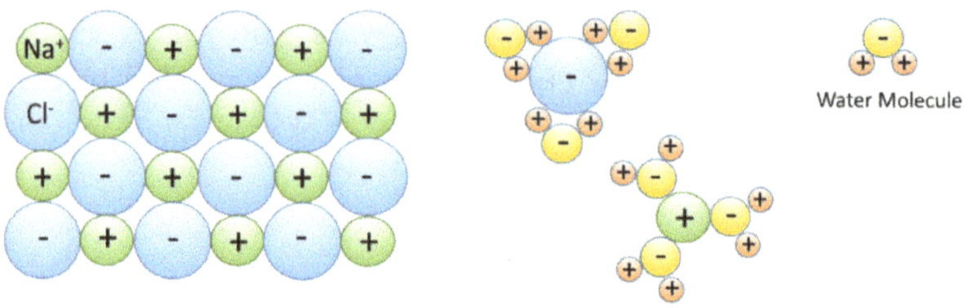

Figure 5.1.3 Attraction between polar water molecules and charged ions (such as in NaCl) is greater than the attraction between the charged ions, causing the ions to dissociate and the salt to dissolve (PW).

The **solid phase is less dense than the liquid phase.** In other words, ice floats. Most substances are denser in the solid form than in the liquid form, as their molecules are more closely packed together as a solid. Water is an exception: the density of fresh water is 1.0 g/cm^3, while the density of ice is 0.92 g/cm^3, and once again, this is due to the action of hydrogen bonds.

As water temperature cools the molecules slow down, eventually slowing enough that hydrogen bonds can form and hold the water molecules in a crystal lattice. The molecules in the lattice are spaced farther apart than the molecules in liquid water, which makes ice less dense than liquid water (Figure 5.1.4). This is familiar to anyone who has ever left a full water bottle in the freezer, only to have it burst as the water freezes and expands.

But the relationship between temperature and water density is not a simple linear one. As water cools, its density increases as expected, as the water molecules slow down and get closer together. However, fresh water reaches its maximum density at a temperature of 4° C, and as it cools beyond that point its density declines as the hydrogen bonds begin to form and the intermolecular spacing increases (Figure 5.1.5 inset). The density continues to decline until the temperature reaches 0° C and ice crystals form, reducing the density dramatically (Figure 5.1.5).

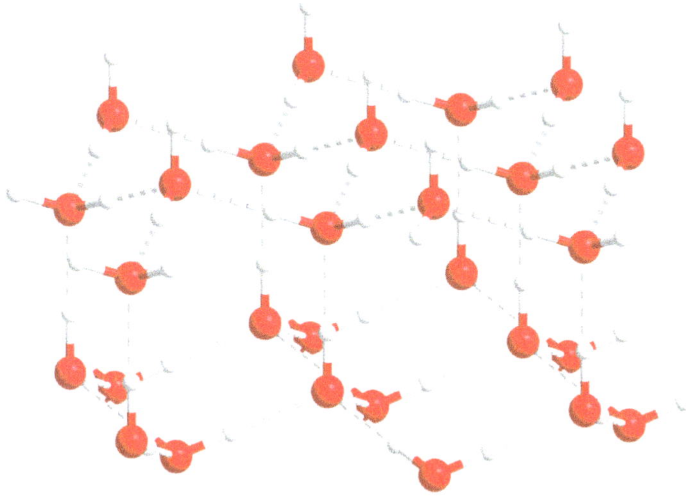

Figure 5.1.4 Crystal lattice structure of ice, showing water molecules held together by hydrogen bonds (By Adam001d (Own work) [CC BY-SA 3.0], via Wikimedia Commons).

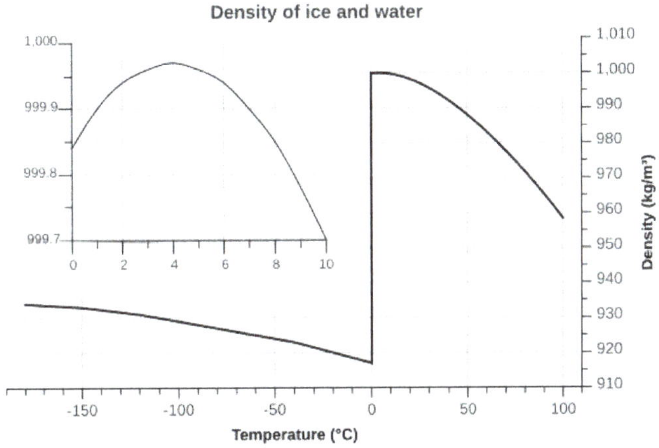

Figure 5.1.5 As temperature declines, the density of water increases until it reaches maximum density at 4° C (inset). Density then declines slightly down to 0° C, where it declines dramatically as ice crystals form (Klaus-Dieter Keller, [CC BY-SA 3.0], via Wikimedia Commons).

There are a number of important implications to ice being less dense than water. Ice floating on the surface of the ocean helps regulate ocean temperatures, and therefore global climate, by influencing the amount of sunlight that is reflected rather than absorbed (see section 8.1). On a smaller scale, surface ice can prevent lakes and ponds from freezing solid during the winter. As fresh surface water cools, the water gets denser, and sinks to the bottom. The new surface water then cools and sinks, and the process is repeated in what is referred to as **overturning**, with denser water sinking and less dense water moving to the surface only to be cooled and sink itself. In this way, the entire body of water is cooled somewhat evenly. This process continues until the surface water cools below 4° C. Below 4° C, the water becomes *less* dense as it cools, so it no longer sinks. Instead, it remains as the surface, getting colder and less dense, until it freezes at 0° C. Once fresh water freezes, the ice floats and insulates the rest of the water beneath it, reducing further cooling. The densest bottom water is still at 4° C, so it does not freeze, allowing the bottom of a lake or pond to remain unfrozen (which is good news for the animals living there) no matter how cold it gets outside.

The dissolved salts in seawater inhibit the formation of the crystal lattice, and therefore make it harder for ice to form. So seawater has a freezing point of about -2° C (depending on salinity), and freezes before a temperature of maximum density is reached. Thus seawater will continue to sink as it gets colder, until it finally freezes.

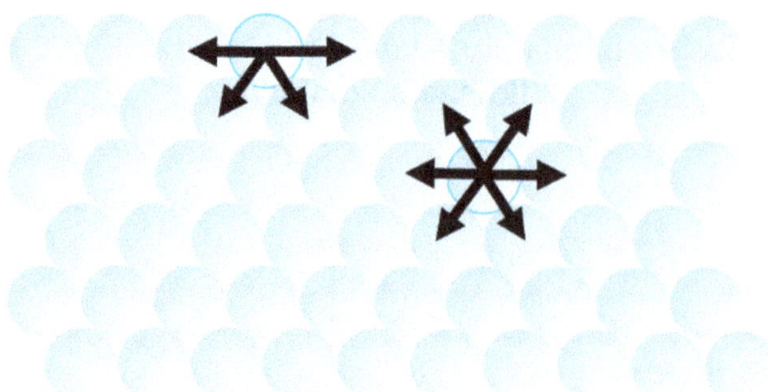

Figure 5.1.6 The net attractive force between molecules at the surface is inwards, leading to surface tension. For molecules in the center, the force is equal in all directions (PW).

Water has a very **high surface tension**, the highest of any liquid except mercury (Table 5.1.2). Water molecules are attracted to each other by hydrogen bonds. For molecules not at the water surface, they are surrounded by other water molecules in all directions, so the attractive forces are evenly distributed in all directions. But for molecules at the surface there are few adjacent molecules above them, only below, so all of the attractive forces are directed inwards, away from the surface (Figure 5.1.6). This inwards force is what causes water droplets to take on a spherical shape, and water to bead up on a surface, as the spherical shape provides the minimum possible surface area. These attractive forces also cause the surface of the water to act like an elastic "skin" which allows things like insects to sit on the water's surface without sinking.

Table 5.1.2 Surface tensions of various liquids

Liquid	Surface Tension (millinewton/meter)	Temperature °C
Mercury	487.00	15
Water	71.97	25
Glycerol	63.00	20
Acetone	23.70	20
Ethanol	22.27	20

5.2 ORIGIN OF THE OCEANS

*Portions modified from Karla Panchuk in "Physical Geology" by Steven Earle**

What is the origin of the oceans? Recall from section 3.1 that the initial formation of the Earth occurred through the accumulation of different substances, and was succeeded by a phase of liquefaction and vigorous volcanic events. The substances that accumulated on the early Earth contained the constituents that would ultimately form our oceans and atmosphere. Several hypotheses exist on the genesis of the oceans. It is suggested that gasses stay dissolved in magma due to the enormous pressures present in the Earth's interior. During volcanic activity, when these magmas ascend to the surface, the pressure decreases, causing the gases to be expelled through a phenomenon known as outgassing. Volcanic eruptions emit a variety of gases, such as water vapor, carbon dioxide (CO2), sulfur dioxide (SO2), carbon monoxide (CO), hydrogen sulfide (H2S), hydrogen gas, nitrogen, and methane (CH4). Gases with lower molecular weights, such as hydrogen and helium, escaped into space, whereas gases with higher molecular weights stayed behind and gave rise to Earth's initial atmosphere, which could have led to the presence of surface water. Another hypothesis suggests that during this initial assault, water was delivered to Earth by comets, which consist primarily of dust and ice, and/or meteorites that potentially carried small amounts of water that may have settled on the Earth's surface. These hypotheses are not mutually exclusive, and it is conceivable that all of them played a role in the creation of the oceans.

The rise of atmospheric oxygen

It's worth noting that the early atmosphere lacked free oxygen (O_2), the form of oxygen that we breathe. We know this in part because prior to 2 billion years ago, there were no sedimentary beds stained red from oxidized iron minerals. Iron minerals were present, but not in oxidized form. At that time, O_2 was produced in the atmosphere when the Sun's ultraviolet rays split water molecules apart; however, chemical reactions removed the oxygen as quickly as it was produced. It wasn't until the appearance of life that Earth's atmosphere began to become oxygenated. Photosynthetic organisms used the abundant CO_2 in the atmosphere to manufacture their food, and released O_2 as a by-product. At first all of the oxygen was consumed by chemical reactions, but eventually the organisms released so much O_2 that it overwhelmed the chemical reactions and oxygen began to accumulate in the atmosphere, although present levels of 21% oxygen didn't occur until about 350 Ma. Today the part of our atmosphere that isn't oxygen consists largely of nitrogen (78%). The oxygen-rich atmosphere on our planet is life's signature. If geologic process were the only processes controlling our atmosphere, it would consist mostly of carbon dioxide, like the atmosphere of Venus.

As the early Earth cooled, the water vapor in the atmosphere condensed and fell as rain. By about 4 billion years ago, the first permanent accumulations of water were present on Earth, forming the oceans and other bodies of water. Water moves between these different reservoirs through the **hydrological cycle**. Water is evaporated from the oceans, lakes, streams, the surface of the land, and plants (transpiration) by solar energy (Figure 5.2.1). It is moved through the atmosphere by winds and condenses to form clouds of water droplets or ice crystals. It comes back down as rain or snow and then flows through streams and rivers, into lakes, and eventually back to the oceans. Water on the surface and in streams and lakes infiltrates the ground to become groundwater. Groundwater slowly moves through the rock and surface materials; some returns to other streams and lakes, and some goes directly back to the oceans.

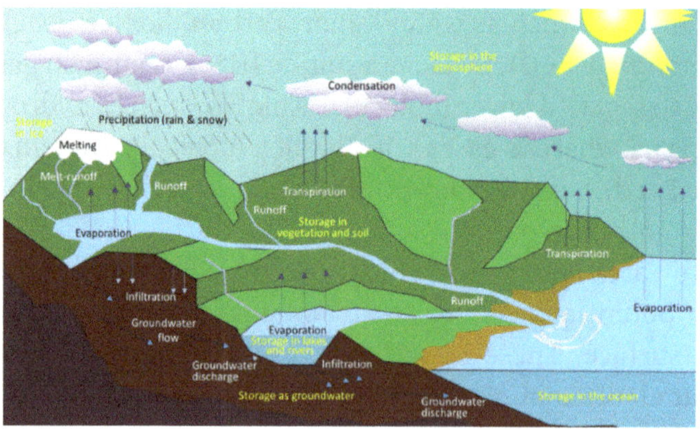

Figure 5.2.1 Earth's hydrological cycle (Steven Earle, "Physical Geology").

Water is stored in various reservoirs as it moves through this cycle. The largest, by far, is the oceans, accounting for 97% of the volume (Figure 5.2.2). Of course, that water is salty. The remaining 3% is fresh water. Two-thirds of our fresh water is stored in ice and one-third is stored as groundwater. The remaining fresh water — about 0.03% of the total — is stored in lakes, streams, vegetation, and the atmosphere.

To put that in perspective, let's think about putting all of Earth's water into a 1 L jug. We start by almost filling the jug with 970 ml of water and 34 g of salt.

Then we add one regular-sized (~20 mL) ice cube (representing glacial ice) and two teaspoons (~10 mL) of groundwater. All of the water that we see around us in lakes and streams and up in the sky can be represented by adding three more drops from an eyedropper.

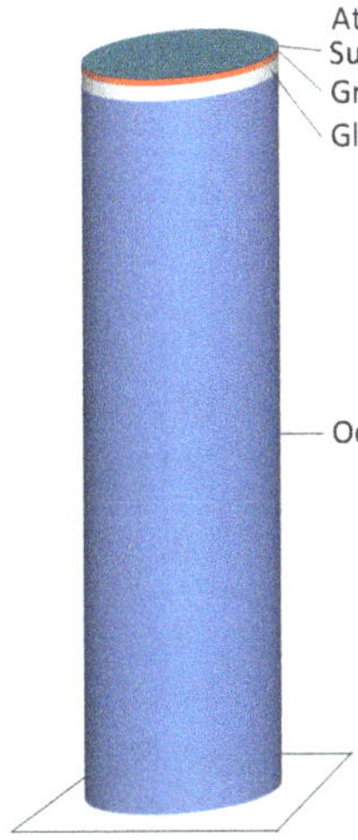

Figure 5.2.2 Proportions of Earth's water found in the various reservoirs (Steven Earle, "Physical Geology").

Although the proportion of Earth's water that is in the atmosphere is tiny, the actual volume is huge. At any given time, there is the equivalent of approximately 13,000 km^3 of water in the air in the form of water vapor and water droplets in clouds. Water is evaporated from the oceans, vegetation, and lakes at a rate of 1,580 km^3 per day, and just about exactly the same volume falls as rain and snow every day, over both the oceans and land. The precipitation that falls on land goes back to the ocean in the form of stream flow (117 km^3/day) and groundwater flow (6 km^3/day).

How did the oceans get salty?

Outgassing was responsible for ocean formation, but how did the ocean water get salty? Most of the salts and dissolved elements in the ocean were probably outgassed along with the water vapor, so the ocean has probably always been about as salty as it is now. But we know that rainfall and other processes weather rocks on the Earth's surface, and runoff carries dissolved substances into the ocean, contributing to its salinity. Yet despite this constant input, the ocean's salt composition remains essentially the same. Therefore, the rate of input of new material must be balanced by the rate of removal; in other words, the oceans are in a **steady state** in regards to salinity.

There are multiple pathways through which dissolved ions enter the ocean; runoff from streams and rivers, volcanic activity, hydrothermal vents (see section

4.11), dissolution or decay of substances in the ocean, and groundwater input. Ions are removed from seawater as they are incorporated by living organisms (for example in shell production) or sediments, sea spray, percolation of water into the crust, or when sea water gets isolated from the ocean and evaporates.

The relationship between the input and removal of an ion can be examined through the concept of **residence time**, which is the average length of time a single atom of an element remains in the ocean before being removed. Residence time is calculated as:

There is great variation in residence times for different substances (Table 5.2.1). Generally speaking, substances that are readily used in biological processes have short residence times, as they are used up as they become available. Substances with longer residence times are less reactive, and may be a part of long-scale geological cycles.

Table 5.2.1 Residence times for some constituents of sea water

Constituent	Residence time (years)
Chloride (Cl^-)	100,000,000
Sodium (Na^+)	68,000,000
Calcium (Ca^{2+})	1,000,000
Water	4100
Iron (Fe)	200

So what about lakes? They are subjected to runoff and river input, so why aren't they salty like the oceans? One reason is that compared to the oceans, lakes and ponds are relatively temporary phenomena, so they do not last long enough to accumulate the same levels of ions as the oceans. Furthermore, lakes often have rivers flowing both into and out of them, so many ions are removed through the outflow, eventually finding their way to the oceans. The oceans only receive river input; there are no rivers flowing out of the ocean to remove these materials, so they are found in greater abundance in sea water. It should be noted that there are some lakes that contain water whose salt content may rival or exceed that of the ocean; these lakes usually lack river outflow. The Great Salt Lake in the western United States is an example.

5.3 SALINITY PATTERNS

All of the salts and ions that dissolve in seawater contribute to its overall **salinity**. Salinity of seawater is usually expressed as the grams of salt per kilogram (1000 g) of seawater. On average, about 35 g of salt is present in each 1 kg of seawater, so we say that the average salinity of the ocean salinity is 35 parts per thousand (ppt). Note that 35 ppt is equivalent to 3.5% (parts per hundred). Some sources now use practical salinity units (PSU) to express salinity values, where 1 PSU = 1 ppt. The units are not included, so we can refer simply to a salinity of 35.

Many different substances are dissolved in the ocean, but six ions comprise about 99.4% of all the dissolved ions in seawater. These six **major ions** are (Table 5.3.1):

Table 5.3.1 The six major ions in seawater

	g/kg in seawater	% of ions by weight
Chloride Cl^-	19.35	55.07%
Sodium Na^+	10.76	30.6%
Sulfate SO_4^{2-}	2.71	7.72%
Magnesium Mg^{2+}	1.29	3.68%
Calcium Ca^{2+}	0.41	1.17%
Potassium K^+	0.39	1.1%
		99.36%

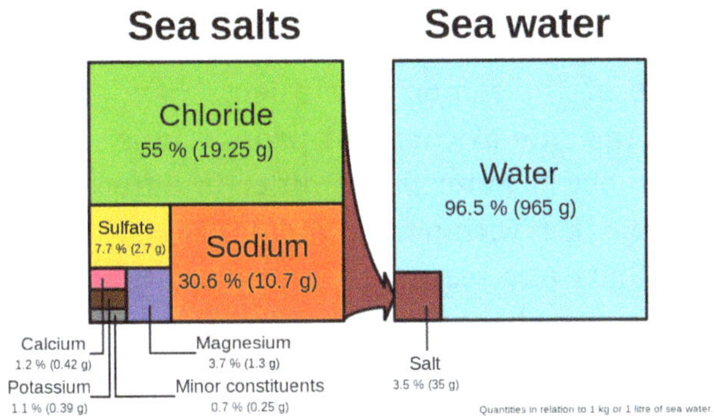

Figure 5.3.1 The relative proportions of ions in seawater. (By derivative work: Tcncv (talk) Sea_salt-e_hg.svg: Hannes Grobe, Alfred Wegener Institute for Polar and Marine Research, Bremerhaven, Germany; SVG version by Stefan Majewsky (Sea_salt-e_hg.svg) [CC BY-SA 2.5], via Wikimedia Commons)

Chloride and sodium, the components of table salt (sodium chloride NaCl), make up over 85% of the ions in the ocean, which is why seawater tastes salty (Figure 5.3.1). In addition to the major constituents, there are numerous minor constituents; radionucleotides, organic compounds, metals etc. These minor constituents are found in concentrations of ppm (parts per million) or ppb (parts per billion), unlike the major ions that are far more abundant (ppt) (Table 5.3.2). To put this into perspective, 1 ppm = 1 mg/kg, or the equivalent of 1 teaspoon of sugar dissolved in 14,000 cans of soda. 1 ppb = 1 μg/kg, or the equivalent of 1 teaspoon of a substance dissolved in five Olympic-sized swimming pools! These minor constituents represent numerous substances, but together they make up less than 1% of the ions in the seawater. Some of these may be important as minerals and trace elements vital to living organisms, but they don't have much impact on the overall salinity. But given the vast size of the oceans, even materials found in trace abundance can represent fairly large reservoirs. For example gold is a trace element in seawater, found in concentrations of parts per trillion, yet if you could extract all of the gold in just one km^3 of seawater, it would be worth about $20 million!

Table 5.3.2 Concentrations of some minor elements in seawater

	g/kg in seawater		g/kg in seawater
Carbon	0.028	Iron	2×10^{-6}
Nitrogen	0.0115	Manganese	2×10^{-7}
Oxygen	0.006	Copper	1×10^{-7}
Silicon	0.002	Mercury	3×10^{-8}
Phosphorous	6×10^{-5}	Gold	4×10^{-9}
Uranium	3.2×10^{-6}	Lead	5×10^{-10}
Aluminum	2×10^{-6}	Radon	6×10^{-19}

Because the six major ions in seawater comprise over 99% of the total salinity, changes in abundance of the minor constituents have little effect on overall salinity. Furthermore, the **rule of constant proportions** states that even though the absolute salinity of ocean water might differ in different places, the relative *proportions* of the six *major ions* within that water are always constant. For example, no matter the total salinity of a seawater sample, 55% of the total salinity will be due to chloride, 30% due to sodium, and so on. Since the proportion of these major ions does not change, we call these **conservative ions.**

Given these constant proportions, in order to calculate total salinity you can simply measure the concentration of just one of the major ions and use that value to calculate the rest. Traditionally chloride has been the ion measured because it is the most abundant, and thus the simplest to measure accurately. Multiplying the concentration of chloride by 1.8 gives the total salinity. For example, looking at Figure 5.3.1, 19.25 g/kg (ppt) chloride x 1.8 = 35 ppt. Today, for rapid measurements of salinity, electrical conductivity is often used rather than determining chloride concentrations (see box below).

Measuring salinity

There are a number of methods available for measuring the salinity of water. The most precise measurements utilize direct chemical analysis of the seawater in a lab setting, but there are a number of ways to get immediate salinity measurements in the field. For a quick estimate of salinity, a hand-held **refractometer** can be used (right).

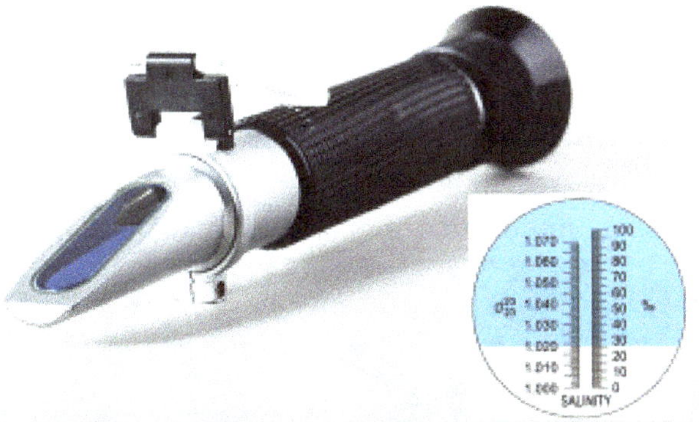

Handheld refractometer (Photo by CEphoto, Uwe Aranas / CC-BY-SA-3.0 via Wikimedia Commons)

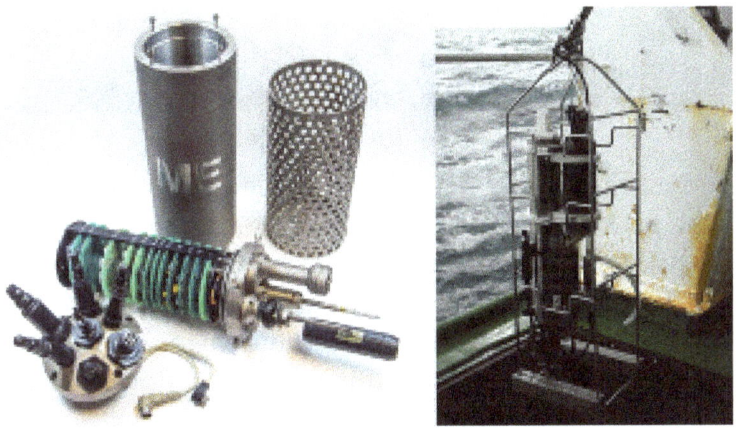

Left: The inner workings of a CTD. An array of different probes are attached. Right: A CTD in a protective cage ready to be deployed at sea. (Left: Hannes Grobe [CC BY 3.0 (https://creativecommons.org/licenses/by/3.0)]; Right: © Hans Hillewaert (CC BY-SA 4.0) via Wikimedia Commons)

This instrument measures the degree of bending, or refraction, of light rays as they pass through a fluid. The greater the amount of dissolved salts in the sample, the greater the degree of light refraction. The observer traps a drop of water on the blue screen, and looks through the eyepiece. The dividing line between the blue and white sections of the scale (inset) can be used to read the salinity.

For more accurate measurements, most oceanographers use an instrument that measures electrical **conductivity**. An electrical current is passed between two electrodes immersed in water, and the higher the salinity, the more readily the current will be conducted (the ions in seawater conduct electrical currents). Conductivity probes are often bundled into an instrument called a **CTD**, which stands for Conductivity, Temperature, and Depth, which are the most commonly-measured parameters. Modern CTDs can be outfitted with an array of probes measuring parameters like light, turbidity (water clarity), dissolved gases etc. CTDs can be large instruments (below), but small hand-held salinity probes are also widely available.

For large-scale salinity measurements, oceanographers can use satellites, such as the Aquarius satellite, which was able to measure surface salinity differences as small as 0.2 PSU as it mapped the ocean surface every seven days (below).

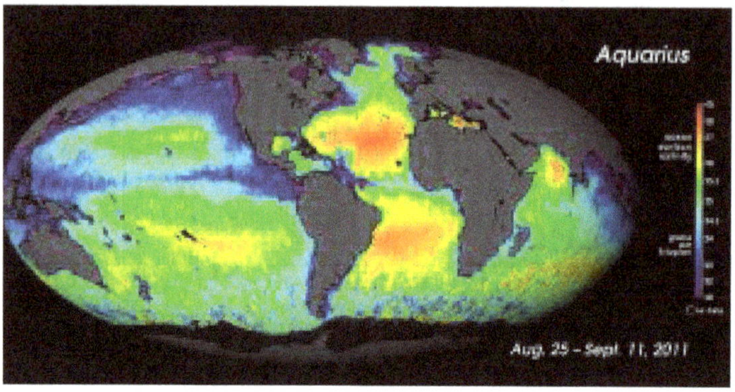

(NASA, Public Domain)

It is important to be aware that while the rule of constant proportions applies to most of the ocean, there may be certain coastal areas where lots of river discharge may alter these proportions slightly. Furthermore, it is important to remember that the rule of constant proportions only applies to the *major* ions. The proportions of the minor ions may fluctuate, but remember that they make a very minor contribution to overall salinity. Because the concentrations of the minor ions are not constant, these are referred to as **non-conservative ions.**

Why are the major ions found in constant proportions? There is constant input of ions from river runoff and other processes, usually in very different proportions from what is found in the ocean. So why don't the proportions in the oceans change? Most of the ions discharged by rivers have fairly low residence times (see section 5.2) compared to ions in seawater, usually because they are used in biological processes. These low residence times do not allow the ions to accumulate and alter salinity. Also, the mixing time of the world ocean is around 1000 years, which is very short compared to the residence times of the major ions, which may be tens of millions of years long. So during the residence time of a single ion the ocean has mixed numerous times, and the major ions have become evenly distributed throughout the ocean.

Variations in Salinity

Total salinity in the open ocean averages 33-37 ppt, but it can vary significantly in different locations. But since the major ion proportions are constant, the regional salinity differences must be due more to water input and removal rather than the addition or removal of ions. Fresh water input comes through processes like precipitation, runoff from land, and melting ice. Fresh water removal primarily comes from evaporation and freezing (when seawater freezes, the resulting ice is mostly fresh water and the salts are excluded, making the remaining water even saltier). So differences in rates of precipitation, evaporation, river discharge, and ice formation play a significant role in regional salinity variations. For example, the Baltic Sea has a very low surface salinity of around 10 ppt, because it is a mostly enclosed body of water with lots of river input. Conversely, the Red Sea is very salty (around 40 ppt), due to the lack of precipitation and the hot environment which leads to high levels of evaporation.

One of the saltiest large bodies of water on Earth is the Dead Sea, between Israel and Jordan. Salinity in the Dead Sea is around 330 ppt, which is almost ten times saltier than the ocean. This extremely high salinity is a result of the hot, arid conditions in the Middle East that lead to high rates of evaporation. In addition, in the 1950s the flow from the Jordan River was diverted away from the Dead Sea, so there is no longer significant fresh water input. With no input and high evaporation, the water level in the Dead Sea is receding at a rate of about 1 m per year. The high salinity makes the water very dense, which creates buoyant forces that allow people to easily float at the surface. But the high salinity also means that the water is too salty for most living organisms, so only microbes are able to call it home; hence the name the Dead Sea. But as salty as the Dead Sea may be, it is not the saltiest body of water on Earth. That distinction currently belongs to Gaet'ale Pond in Ethiopia, with a salinity of 433 ppt!

Latitudinal Variations

While local conditions are important for determining salinity patterns in any single location, there are some global patterns that bear further investigation. Temperature is highest at the equator, and lowest near the poles, so we would expect higher rates of evaporation, and therefore higher salinity, in equatorial regions (Figure 5.3.2). This is generally the case, but in the figure below salinity right along the equator seems to be a little lower than at slightly higher latitudes. This is because equatorial regions also get a high volume of rain on a regular basis, which dilutes the surface water along the equator. So the higher salinities are found at subtropical, warm latitudes with high evaporation and less precipitation. At the poles there is little evaporation, which, coupled with ice and snow melting, produces a relatively low surface salinity. The image below shows high salinity in the Mediterranean Sea; this is located in a warm region with high evaporation, and the sea is largely isolated from mixing with the rest of the North Atlantic water, leading to high salinity. Lower salinities, such as those around southeast Asia, are the result of precipitation and high volumes of river input.

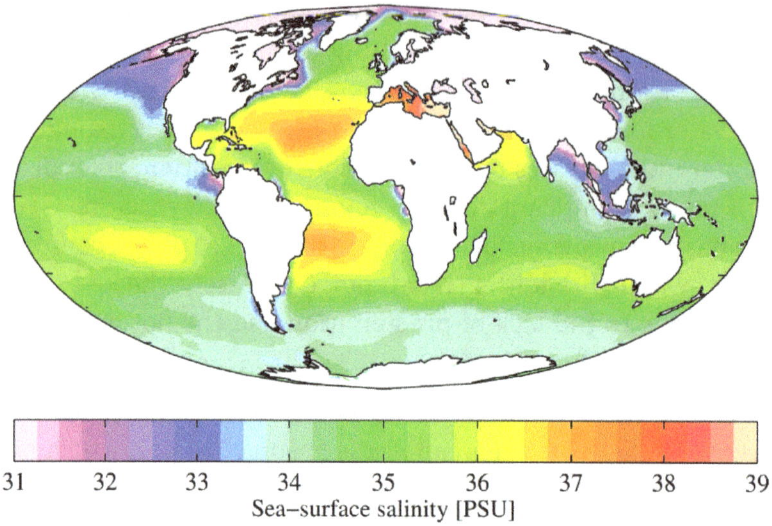

Figure 5.3.2 Annual mean global sea surface salinity (Data from World Ocean Atlas 2009. By Plumbago – Own work, [CC BY-SA 3.0], via Wikimedia Commons).

Figure 5.3.3 shows the mean global differences between evaporation and precipitation (evaporation – precipitation). Green colors represent areas where precipitation exceeds evaporation, while brown regions are where evaporation is greater than precipitations. Note the correlation between precipitation, evaporation, and surface salinity as seen in Figure 5.3.2.

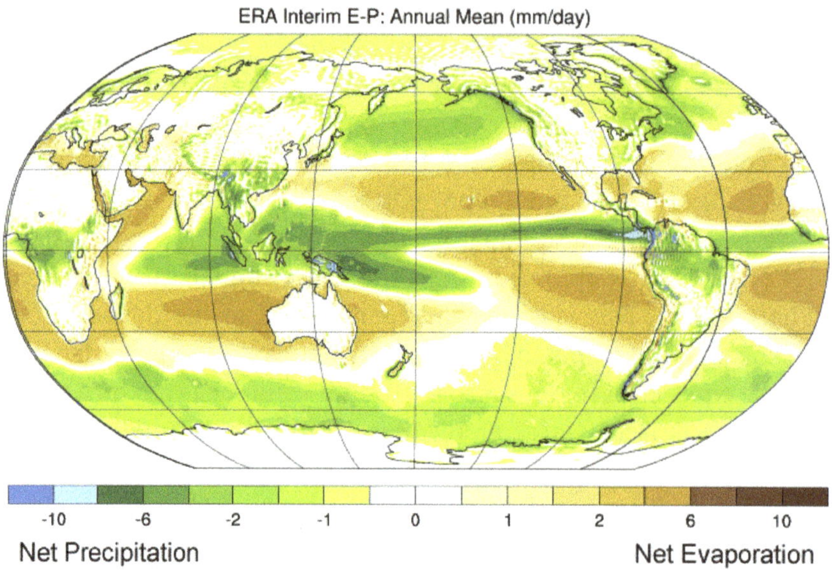

Figure 5.3.3 Annual mean global differences in evaporation and precipitation (J.Fasullo. National Center for Atmospheric Research Staff (Eds). "The Climate Data Guide: ERA-Interim: derived components." Retrieved from https://climatedataguide.ucar.edu/climate-data/era-interim-derived-components).

Vertical Variation

Furthermore, salinity exhibits variations not just across different geographical locations but also inside different depths. The primary factors contributing to variances in salinity are evaporation, precipitation, runoff, and ice cover. All of these processes take place near the surface of the ocean, rather than at deeper levels. Therefore, the most significant variations in salinity are expected to be observed in the surface waters. The salinity in deeper water is consistently uniform, as it is not influenced by these surface processes. Figure 5.3.4 displays many salinity profiles that serve as exemplary examples. The uppermost layer of the ocean, known as the mixed layer, exhibits a rather consistent level of salinity over approximately the first 200 meters below the surface. The movement of winds, waves, and surface currents agitate the surface water, resulting in extensive mixing within this layer and rather consistent salinity conditions. Below the mixed layer is a zone characterized by a fast variation in salinity over a relatively modest change in depth. The area of significant and swift transformation is referred to as the halocline, which serves as a boundary between the mixed layer and the deep ocean. Below the halocline, the salinity remains relatively constant all the way to the bottom, as this area is not influenced by surface processes that affect salinity. Observe the reduced salinity on the surface of the Earth at high latitudes, and the increased salinity at low latitudes, as previously said. However, despite the apparent disparities, the salinity at deeper levels in both areas may exhibit significant similarities.

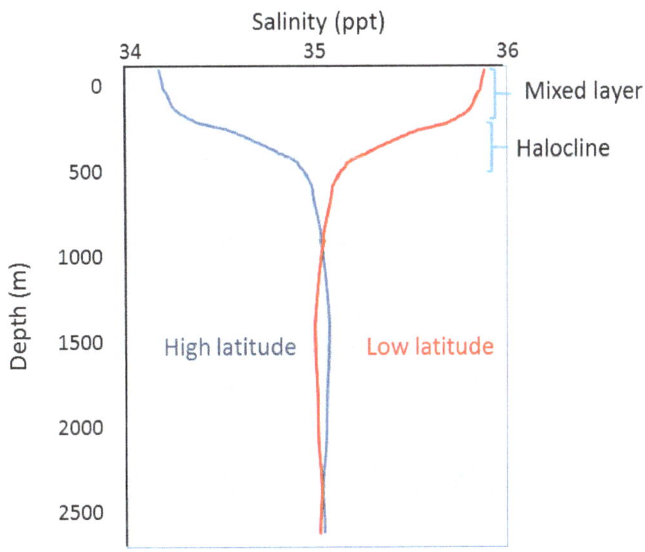

Figure 5.3.4 Salinity profiles from two hypothetical sites in the open ocean, one from high latitude and one from low latitude (PW).

5.4 DISSOLVED GASES: OXYGEN

Seawater contains dissolved compounds other than ions. The oceans also harbor dissolved gases that are crucial for living organisms, notably oxygen (O2), carbon dioxide (CO2), and nitrogen (N2). Marine plants, algae, and phytoplankton (known as primary producers) as well as animals rely on oxygen for respiration. Primary producers utilize carbon dioxide to fuel the process of photosynthesis, which results in the production of oxygen as a byproduct. Bacteria in the ocean fix dissolved nitrogen gas, converting it into nitrate and nitrite forms that are essential for primary production.

All of these gases are present in the atmosphere and can permeate the ocean by undergoing dissolution into the water near the ocean's surface. However, the concentration of each gas in the atmosphere significantly differs from its concentration in the ocean (Table 5.4.1).

Table 5.4.1 Percentage of total gas in each compartment

	Air	Total Ocean	Surface Ocean
N_2	78%	11%	48%
O_2	21%	6%	36%
CO_2	0.04%	83%	15%

The solubility and saturation of a gas in water determine the quantity of that gas that can dissolve in the ocean. Solubility is the maximum quantity of a gas that can be dissolved in water under specific conditions, typically at a temperature of 0o C and a pressure of 1 atmosphere. The solubility of a gas is positively correlated with higher pressure, lower temperature, and lower salinity. Saturation pertains to the current level of gas dissolved in the water, in relation to the maximal capacity. When the water is not fully saturated, it has the ability to dissolve additional gas. Gas may be released if the water is saturated or supersaturated. The majority of atmospheric gases in the ocean are saturated, with the exception of O2 and CO2, which are not saturated due to their quick utilization by living organisms.

Oxygen

Typical oceanic dissolved oxygen profiles are shown in Figure 5.4.1. The shape of the profile is determined by the various processes that add or remove oxygen from the water at different depths.

Oxygen content is highest at the surface for two main reasons; this is where oxygen dissolves into the ocean from the atmosphere, and the surface water is where oxygen is produced by phytoplankton through photosynthesis. Respiration is also occurring in the surface waters, but the rate of photosynthetic oxygen production is greater than the rate of removal through respiration. It should be noted that even though dissolved oxygen is highest at the surface, there is still far less oxygen in the water than is found in the air. Well-oxygenated surface water may only contain around 8 mg O_2/l, while the air contains 210 mg O_2/l.

As depth increases, dissolved oxygen declines, reaching a minimum between a few hundred meters and 1000 m deep, the aptly-named **oxygen minimum layer**. At these depths and below, the water is too far removed from the surface for any atmospheric exchange, and there is not enough light to support photosynthesis, so there is little if any oxygen added to the water. At the same time, oxygen is removed from the water through the respiration of deep water organisms, and the decomposition of organic material by bacteria as it sinks to depth.

Below the oxygen minimum layer there is often an increase in dissolved oxygen at the greatest depths (Figures 5.4.1, 5.4.2). This bottom water is usually colder than the surface water and is under enormous pressure; as stated above, lower temperatures and higher pressure increase the solubility of dissolved gases. But there is another reason that bottom water contains more oxygen than mid-water depths that has to do with the way water circulates throughout the deep ocean (see section 9.8). In polar regions, the cold surface water absorbs lots of oxygen. This cold, oxygen-rich water sinks to the bottom due to its high density, taking the oxygen with it. The oxygen-rich bottom water will then spend the next thousand years or so moving over the seafloor throughout the major ocean basins. This deep water circulation is the source of oxygen for bottom-dwelling (benthic) organisms. The oxygen-rich bottom water forms in the polar regions of the Atlantic, and slowly makes its way to the Pacific, with oxygen being removed for respiration along the way. This is why dissolved oxygen levels in Pacific deep water are generally lower than in the Atlantic (Figure 5.4.1).

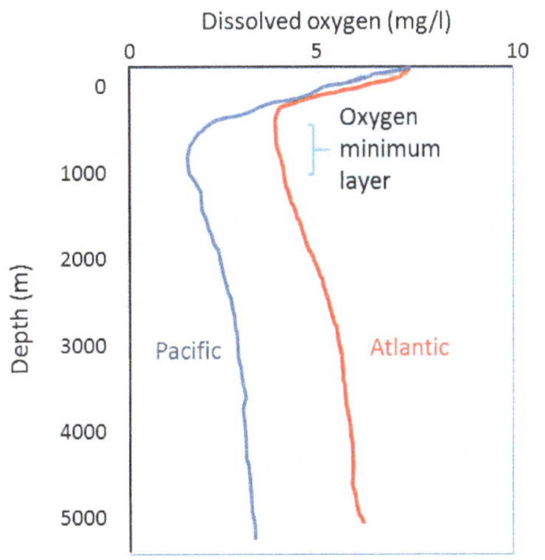

Figure 5.4.1 Representative dissolved oxygen profiles for the Pacific and Atlantic oceans (PW).

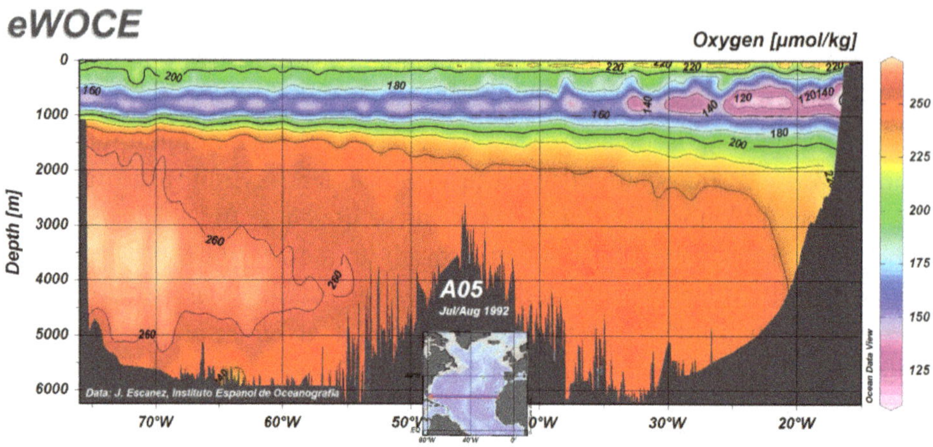

Figure 5.4.2 Dissolved oxygen profile from a transect across the Atlantic Ocean from Florida to the coast of Africa (inset). The oxygen minimum layer is visible between 500-1000 m (eWOCE, http://www.ewoce.org/gallery/eWOCE_Tables.html#Atlantic).

Areas where dissolved oxygen levels are too low to support most life are referred to as **hypoxic** zones (they are experiencing **hypoxia**, or low oxygen). Hypoxia is usually defined as oxygen levels below 2 mg/L. **Anoxic** zones (anoxia = without oxygen) show more severe forms of hypoxia, with oxygen below 0.5 mg/L. Some parts of the oceans may experience seasonal or temporary periods of hypoxia, while in other areas these conditions may last much longer. These hypoxic conditions often lead to mass die-offs of marine organisms who struggle to survive without sufficient oxygen.

5.5 DISSOLVED GASES: CARBON DIOXIDE, PH, AND OCEAN ACIDIFICATION

Oxygen and carbon dioxide are involved in the same biological processes in the ocean, but in opposite ways; photosynthesis consumes CO_2 and produces O_2, while respiration and decomposition consume O_2 and produce CO_2. Therefore it should not be surprising that oceanic CO_2 profiles are essentially the opposite of dissolved oxygen profiles (Figure 5.5.1). At the surface, photosynthesis consumes CO_2 so CO_2 levels remain relatively low. In addition, organisms that utilize carbonate in their shells are common near the surface, further reducing the amount of dissolved CO_2.

In deeper water, CO_2 concentration increases as respiration exceeds photosynthesis, and decomposition of organic matter adds additional CO_2 to the water. As with oxygen, there is often more CO_2 at depth because cold bottom water holds more dissolved gases, and high pressures increase solubility. Deep water in the Pacific contains more CO_2 than the Atlantic as the Pacific water is older and has accumulated more CO_2 from the respiration of benthic organisms.

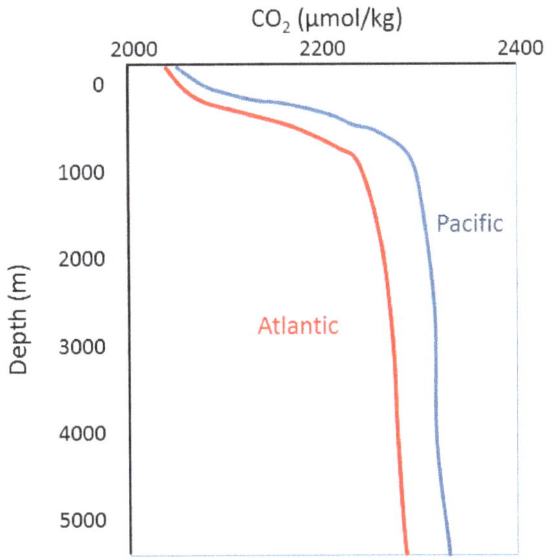

Figure 5.5.1 Representative carbon dioxide profiles for the Pacific and Atlantic oceans (PW).

But the behavior of carbon dioxide in the ocean is more complex than the figure above would suggest. When CO_2 gas dissolves in the ocean, it interacts with the water to produce a number of different compounds according to the reaction below:

$$CO_2 + H_2O \leftrightarrow H_2CO_3 \leftrightarrow H^+ + HCO_3^- \leftrightarrow 2H^+ + CO_3^{2-}$$

CO_2 reacts with water to produce carbonic acid (H_2CO_3), which then dissociates into bicarbonate (HCO_3^-) and hydrogen ions (H^+). The bicarbonate ions can further dissociate into carbonate (CO_3^{2-}) and additional hydrogen ions (Figure 5.5.2).

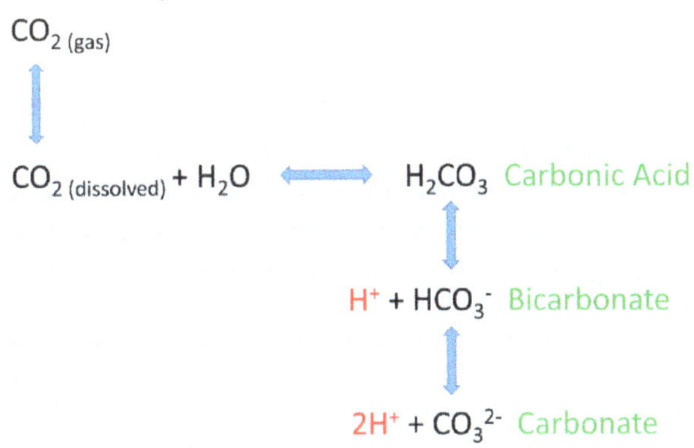

Figure 5.5.2 The fate of dissolved carbon dioxide in the oceans. Most of the carbon ends up in the form of bicarbonate (PW).

Most of the CO_2 dissolving or produced in the ocean is quickly converted to bicarbonate. Bicarbonate accounts for about 92% of the CO_2 dissolved in the ocean, and carbonate represents around 7%, so only about 1% remains as CO_2, and little gets absorbed back into the air. The rapid conversion of CO_2 into other forms prevents it from reaching equilibrium with the atmosphere, and in this way, water can hold 50-60 times as much CO_2 and its derivatives as the air.

CO_2 and pH

The equation above also illustrates carbon dioxide's role as a buffer, regulating the pH of the ocean. Recall that pH reflects the acidity or basicity of a solution. The pH scale runs from 0-14, with 0 indicating a very strong acid, and 14 representing highly basic conditions. A solution with a pH of 7 is considered neutral, as is the case for pure water. The pH value is calculated as the negative logarithm of the hydrogen ion concentration according to the equation:

$$pH = -\log_{10}[H^+]$$

Therefore, a high concentration of H^+ ions leads to a low pH and acidic condition, while a low H^+ concentration indicates a high pH and basic conditions. It should also be noted that pH is described on a logarithmic scale, so every one point change on the pH scale actually represents an order of magnitude (10 x) change in solution strength. So a pH of 6 is 10 times more acidic than a pH of 7, and a pH of 5 is 100 times (10 x 10) more acidic than a pH of 7.

Carbon dioxide and the other carbon compounds listed above play an important role in buffering the pH of the ocean. Currently, the average pH for the global ocean is about 8.1, meaning seawater is slightly basic. Because most of the inorganic carbon dissolved in the ocean exists in the form of bicarbonate, bicarbonate can respond to disturbances in pH by releasing or incorporating hydrogen ions into the various carbon compounds. If pH rises (low $[H^+]$), bicarbonate may dissociate into carbonate, and release more H^+ ions, thus lowering pH. Conversely, if pH gets too low (high $[H^+]$), bicarbonate and carbonate may incorporate some of those H^+ ions and produce bicarbonate, carbonic acid, or CO_2 to remove H^+ ions and raise the pH. By shuttling H^+ ions back and forth between the various compounds in this equation, the pH of the ocean is regulated and conditions remain favorable for life.

CO_2 and Ocean Acidification

In recent years there has been rising concern about the phenomenon of **ocean acidification**. As described in the processes above, the addition of CO_2 to seawater lowers the pH of the water. As anthropogenic sources of atmospheric CO_2 have increased since the Industrial Revolution, the oceans have been absorbing an increasing amount of CO_2, and researchers have documented a decline in ocean pH from about 8.2 to 8.1 in the last century. This may not appear to be much of a change, but remember that since pH is on a logarithmic scale, this decline represents a 30% increase in acidity. It should be noted that even at a pH of 8.1 the ocean is not actually acidic; the term "acidification" refers to the fact that the pH is becoming lower, i.e. the water is moving towards more acidic conditions.

Figure 5.5.3 presents data from observation stations in and around the Hawaiian Islands. As atmospheric levels of CO_2 have increased, the CO_2 content of the ocean water has also increased, leading to a reduction in seawater pH. Some models suggest that at the current rate of CO_2 addition to the atmosphere, by 2100 ocean pH may be further reduced to around 7.8, which would represent more than a 120% increase in ocean acidity since the Industrial Revolution.

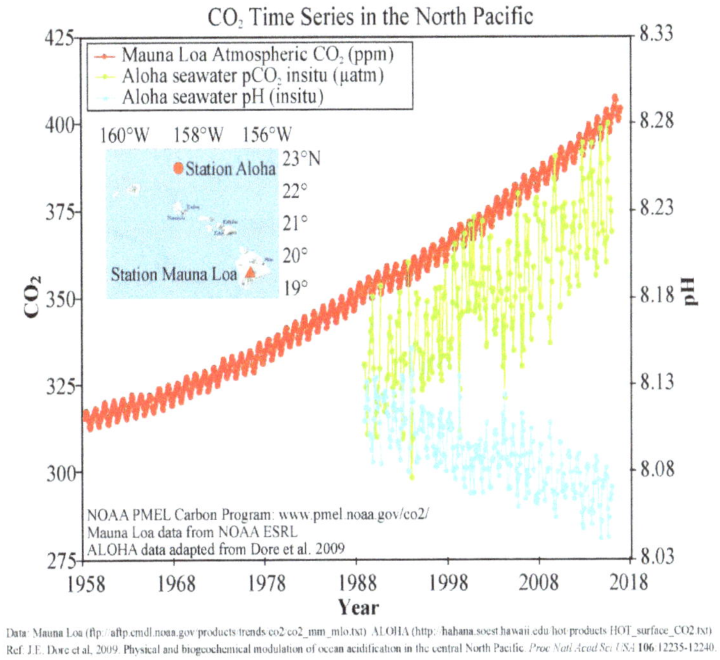

Figure 5.5.3 Changes in atmospheric CO_2 (red), seawater CO_2 (green) and pH (blue) in the Hawaiian Islands (NOAA PMEL).

Why is this important? Declining pH can impact many biological systems. Of particular concern are organisms that secrete calcium carbonate shells or skeletons, such as corals, shellfish, and may planktonic organisms. At lower pH levels, calcium carbonate dissolves, eroding the shells and skeletons of these organisms (Figure 5.5.4).

Not only does a declining pH lead to increased rates of dissolution of calcium carbonate, it also diminishes the amount of free carbonate ions in the water. The relative proportions of the different carbon compounds in seawater is dependent on pH (Figure 5.5.6). As pH declines, the amount of carbonate declines, so there is less available for organisms to incorporate into their shells and skeletons. So ocean acidification both dissolves existing shells and makes it harder for shell formation to occur.

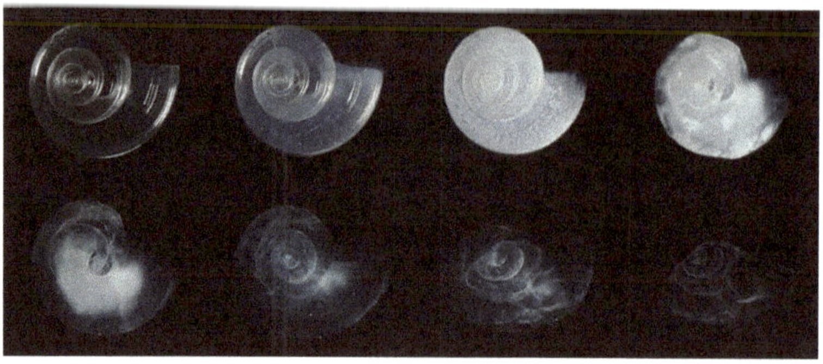

Figure 5.5.5 The results of an experiment placing the calcium carbonate shells of pterapods in seawater with a pH of 7.8, the projected ocean pH for the year 2100 under current rates of acidification. The top row shows the shells before the experiment, and the bottom row shows the dissolution of the shells after 45 days of exposure (NOAA).

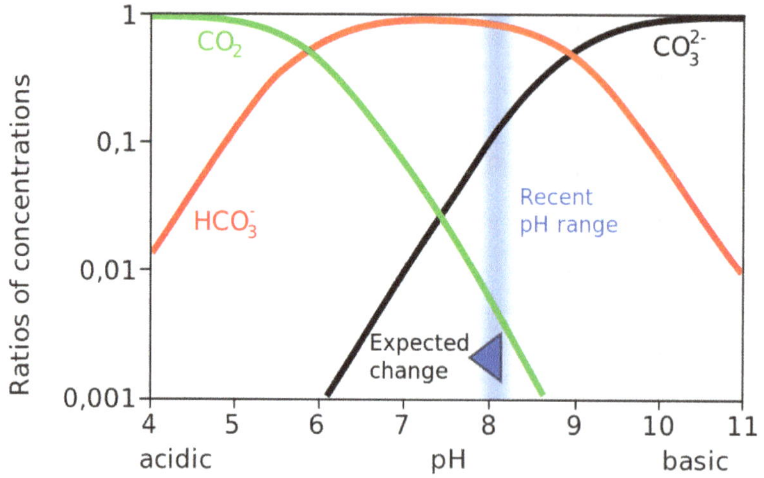

Figure 5.5.6 Proportions of carbon compounds in the ocean at various pH levels. As the ocean pH declines, the proportion of carbonate ions also declines, reducing rates of shell formation (NOAA).

Additional links for more information:

- NOAA Ocean Acidification Program website http://oceanacidification.noaa.gov/

5.6 NITROGEN AND NUTRIENTS

Nitrogen is the most abundant gas in the atmosphere, and like the other atmospheric gases it dissolves into the surface layers of the ocean. But most marine organisms cannot directly utilize dissolved nitrogen in the form in which it exists in air (N_2), so it must first be converted into other nitrogenous products by marine bacteria (Figure 5.6.1). Some bacteria (cyanobacteria) take the dissolved N_2 and convert it into ammonium (NH_4^+) through nitrogen fixation. Some of this ammonium can be used directly by phytoplankton, but the majority of it is converted by bacteria into nitrite (NO_2^{2-}) or nitrate (NO_3^-) through the process of nitrification. Nitrate is the main nitrogenous compound utilized by primary producers in the ocean; it is a major **nutrient** required for photosynthesis. Note that in this context, a nutrient refers to a chemical needed to support photosynthesis and primary production. It does not refer to the nutritional needs of consumer organisms. The nitrogen taken in by phytoplankton gets passed on to consumer organisms, and then gets returned to the ocean through decomposition of wastes and organic matter as these organisms die and sink into deeper water. Finally, the ammonium, nitrate and nitrite can undergo denitrification by yet another group of bacteria and get converted back into N_2, which can reenter the cycle or be exchanged with the atmosphere (Figure 5.6.1).

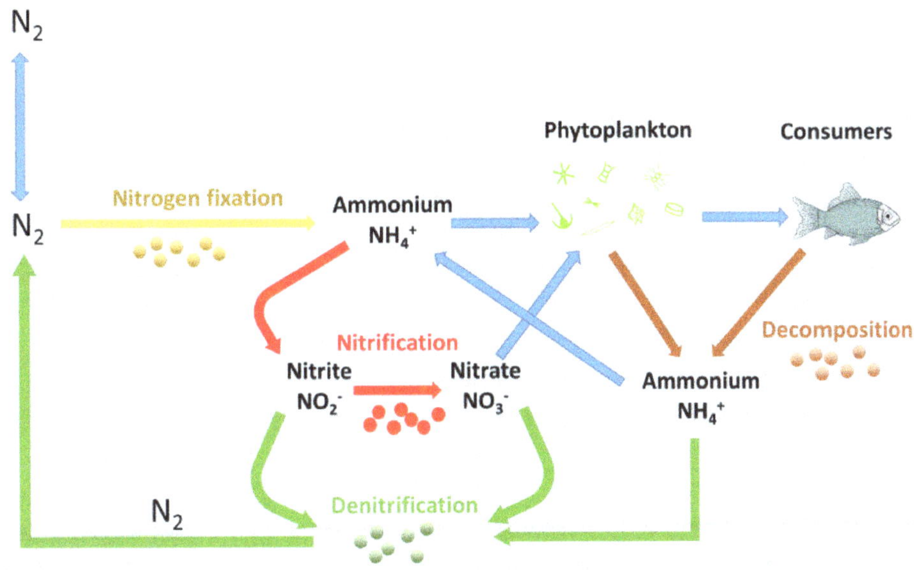

Figure 5.6.1 Simplified nitrogen cycle in the ocean. Colored dots represent the marine bacteria responsible for nitrogen cycling (PW).

Since nitrate is one of the most important nutrients, for now we will focus only on nitrate as we discuss general nutrient patterns in the ocean. A representative nutrient (nitrate) profile is shown below (Figure 5.6.2). Since nutrients are rapidly used in biological processes, they are non-conservative, and their concentrations vary regionally and seasonally. Nutrient concentrations are low at the surface, because that is where the primary producers are located; the nutrients are rapidly consumed and they do not have the chance to accumulate. Nutrient levels increase at depth, as they are no longer being consumed by producers, and they are being regenerated through the decomposition of organic material by bacteria.

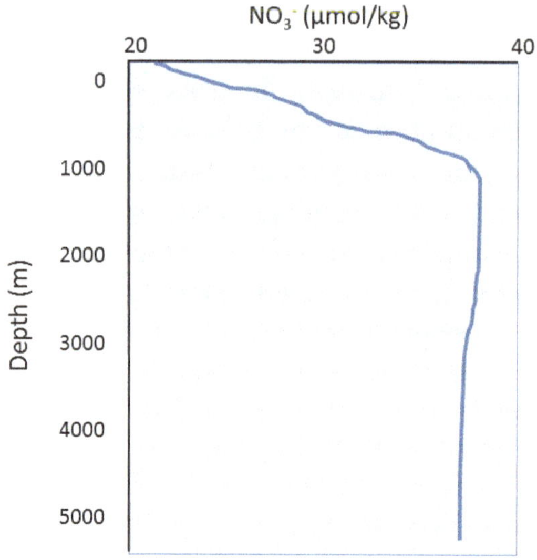

Figure 5.6.2 Representative nutrient (nitrate) profile for the open ocean (PW).

Comparisons of nutrient and dissolved oxygen profiles between the Pacific and Atlantic Oceans reveal some interesting differences (Figure 5.6.3). In general, the Atlantic has more dissolved oxygen but lower nutrient concentrations than the Pacific. Water masses form in the North Atlantic that are very cold and dense, so the water sinks to the bottom. This water will then spend the next thousand years or more moving along the seafloor from the Atlantic, to the Indian, and finally into the Pacific Ocean (see section 9.8).

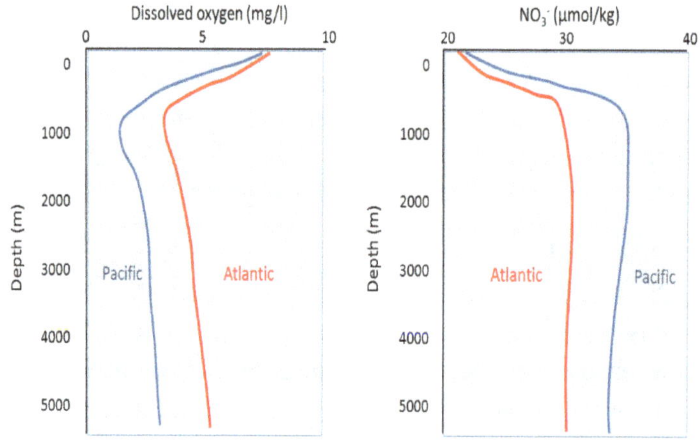

Figure 5.6.3 Dissolved oxygen (left) and nutrient (right) profiles for the Pacific and Atlantic Oceans. As water circulates from the Atlantic to the Pacific, oxygen is consumed while nutrients accumulate (PW).

This water is initially oxygen-rich surface water, and as it sinks it brings oxygen to the deep seafloor. As the bottom water moves across the ocean basins, oxygen is removed through respiration and decomposition, and by the time it arrives in the Pacific it has been depleted of much of its oxygen. At the same time, decomposition of sinking organic matter adds nutrients to the deep water as it moves through the oceans, so nutrients accumulate and the Pacific water becomes nutrient-rich. Comparison of the ratios of oxygen to nutrients in the deep water can therefore provide an indication of the age of the water, i.e. how much time has passed since it initially sank from the surface in the North Atlantic. Water with a high oxygen and low nutrient content is relatively young, while older water will have less oxygen but higher nutrient concentrations.

Other important nutrients, such as phosphate and silica, show similar patterns to nitrate (Figure 5.6.4), and will be discussed in the section on primary production.

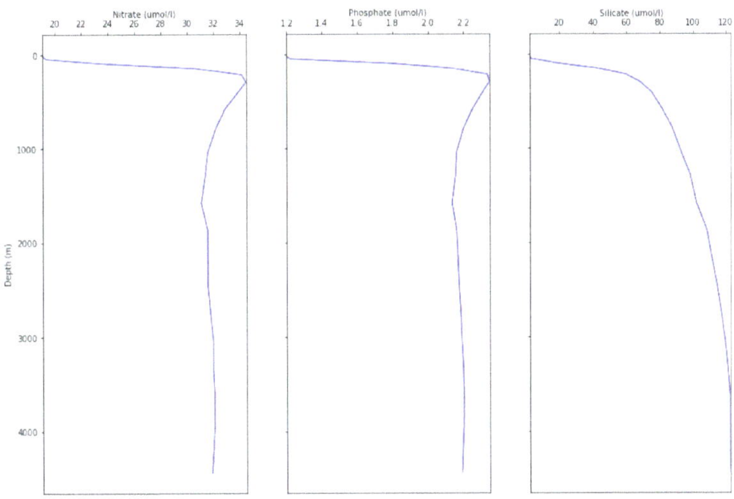

Figure 5.6.4 Nitrate, phosphate, and silicate profiles from an open-ocean location in the South Atlantic (52° S, 35°13'58.8″ W), north of South Georgia Island (image by PW, data from 2014, World Ocean Database).

5.7 CLASSIFYING ELEMENTS IN SEAWATER

Having gained knowledge on the plentiful presence and characteristics of some dissolved compounds in the water, we can now attempt to categorize these elements according to their dispersion. Dissolved substances are typically categorized into various groups:

- Conservative Elements refer to substances that maintain a consistent concentration throughout the ocean, without significant variation in both vertical and horizontal directions. This group include the primary ions found in saltwater, such as sodium or chloride, which have extended periods of time in the ocean and their concentration only fluctuates due to the introduction or removal of freshwater (see to section 5.3).

- The distribution of Nutrient-like Elements closely resembles the profile we previously defined for nitrate. The concentrations of these chemicals are minimal near the surface due to their quick depletion by biological activities. Concentrations of these compounds will increase below the photic zone, as microorganisms decompose them and cycle them back into the water column.
- Scavenged elements are particles that undergo a reaction with other particles and become attached to the surface of the particle. When the particles descend, those elements are extracted and deposited into the sediment. Profiles of scavenged materials often exhibit greater prevalence near the ocean's surface, where the substances enter, but diminish in quantity as depth increases due to their removal by sinking particles. This profile is typical for metals, such as aluminum or lead.
- Inert gases dissolve into the ocean from the atmosphere. Due to their low reactivity, the ocean water becomes saturated with these gases. Due to the phenomenon of increased solubility of gases in colder water, deep, cold water tends to contain higher quantities of certain gases, while warmer surface water contains lower concentrations of them.

6 Chapter

PHYSICAL OCEANOGRAPHY: PHYSICAL PROPERTIES AND PROCESSES OF THE OCEAN

6.1 PRESSURE

When discussing pressure in the ocean, we are specifically referring to hydrostatic pressure. This pressure arises from the gravitational force exerted by the weight of the water column, which presses down on an object. As you descend further, the amount of water above you increases, resulting in a greater weight and hence more pressure. At the Earth's surface, we encounter a pressure of one atmosphere (1 atm = 101.3 kPa) as a result of the gravitational force exerted by the atmosphere above us. As you go deeper into the ocean, the pressure increases in a linear manner with the depth. Specifically, the pressure increases by 1 atmosphere for every 10 meters increase in depth. At a depth of 1000 meters, the pressure would be 101 atmospheres, which is the sum of the 100 atmospheres of pressure resulting from the depth and the 1 atmosphere already existent at the surface. Consequently, the profound regions of the ocean experience extremely elevated pressure. To illustrate, the average depth of the ocean is approximately 3800 m, resulting in a pressure at that depth that is 381 times more than the pressure at the surface. The ramifications of this immense pressure are evident in the video provided, which depicts the drilling of a hole in a pipe located at a depth of 6 miles (about 9.6 km). The external pressure exerted on the pipe exceeds 960 atm, far surpassing the inside pressure. This disparity in pressure has significant consequences for a crab that happens to be traveling by.

High pressure at depth has several significant impacts. Boyle's Law asserts that the volume of a gas is inversely proportional to its pressure. Therefore, high pressure will cause the compression of air spaces, such as the lungs of a diving animal or person, or the interior of a submarine. Submarines and submersibles require exceptionally sturdy hulls to withstand the immense pressure experienced at great depths. Furthermore, according to Henry's Law, an increase in pressure results in a greater amount of gas being dissolved in a fluid. According to the information presented in section 5.4, it may be inferred that water at greater depths and under higher pressure is likely to have a higher concentration of dissolved gases compared to water at the surface.

Furthermore, this also has ramifications for individuals who engage in diving activities. According to Henry's Law, a rise in pressure results in a proportional increase in the solubility of a gas in a fluid, such as blood. In contrast, decreasing pressure causes the fluid to retain a smaller amount of dissolved gas, resulting in the expulsion of surplus gas from the solution, typically in the form of bubbles. When a bottle of a carbonated beverage is opened, the following events occur. The bottle contains a pressurized fluid that, when opened, releases the pressure and allows the dissolved CO2 to escape, resulting in the formation of bubbles. Decompression sickness, also known as "the bends," is a condition that affects SCUBA divers who ascend too rapidly after inhaling compressed air. A gradual ascent enables the elimination of this surplus gas from the bloodstream through exhalation. However, if the diver ascends rapidly, these gases will separate from the solution and create bubbles in the blood, which gather around the joints, resulting in severe pain and potentially fatal consequences.

6.2 TEMPERATURE

Generally ocean temperatures range from about -2° to 30° C (28-86° F). The warmest water tends to be surface water in low latitude regions, while the surface water at the poles is obviously much colder (Figure 6.2.1). Note that at equivalent latitudes, water on the eastern side of the ocean basins is colder than the water on the western side. This has to do with the pattern of surface currents, as described in section 9.1. Even though surface water can be quite warm, most of the water in the oceans is deeper, colder water, so that the average temperature of the entire ocean is about 4° C, which is roughly the temperature inside your refrigerator.

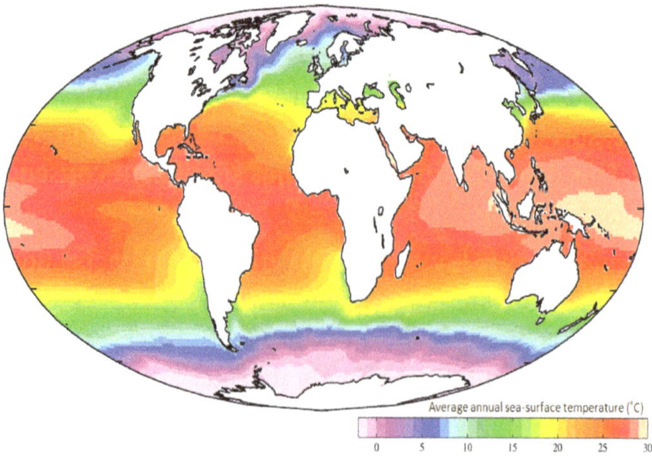

Figure 6.2.1 Global average annual sea surface temperature (Stephen Earle, "Physical Geology").

A typical temperature profile for open ocean, mid-latitude water is shown in Figure 6.2.2. Water is warmest at the surface, as it is warmed by the sun, and the sun's rays can only penetrate depths less than 1000 m (section 6.5). Since the surface water is warmer it is also less dense than the deep water (section 6.3), so it remains at the surface where it can be warmed even more. Temperature is fairly constant in the upper 100-200 m in what is called the **mixed layer**. The mixed layer results from surface winds, waves, and currents that mix the upper water and distribute the heat throughout this layer. Below the mixed layer there is a rapid decline in temperature over a fairly narrow increase in depth. This is called the **thermocline.** Below the thermocline the deep ocean temperature is fairly constant at about 2° C, continuing down to the bottom. There is little temperature change in the deep ocean, as it is far removed from significant heat sources, making it one of the most thermally stable regions on earth. Temperature may fluctuate by less than half a degree per year in the deep ocean (Figure 6.2.3).

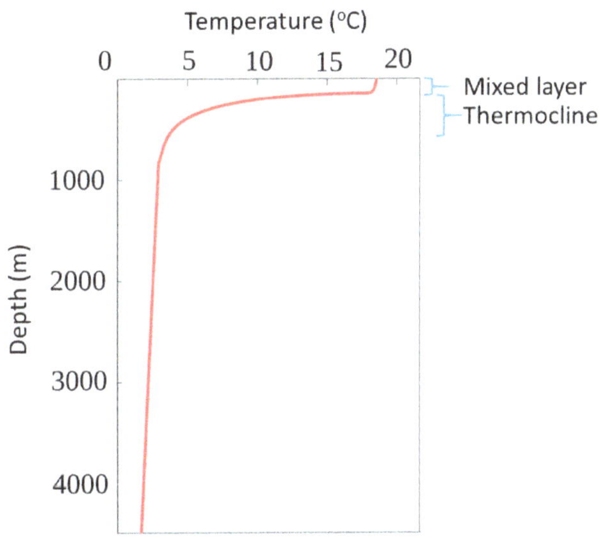

Figure 6.2.2 Typical open ocean temperature profile for a mid-latitude region, showing the mixed layer, steep thermocline, and relatively stable temperature at depth (Public domain via Wikimedia Commons).

Temperature profiles vary at different latitudes, as the surface water is warmer near the equator and colder at the poles. In low latitude tropical regions the sea surface is much warmer, leading to a highly pronounced thermocline (Figure 6.2.4). Additionally, there is not much seasonal change in surface temperature in tropical regions, so there is little seasonal change in the profiles. In high latitude (polar) regions, there is little difference between the surface temperature and the deep-water temperature, and temperature is fairly constant (and cold) at all depths. Polar waters therefore lack a strong thermocline, and as with tropical water, there is little seasonal change in temperatures.

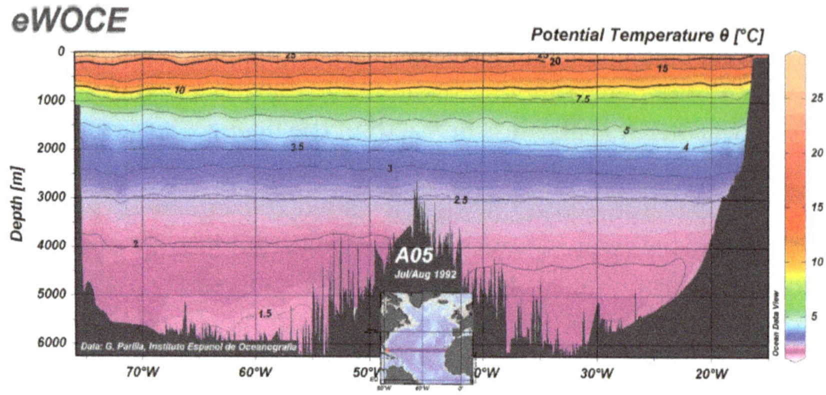

Figure 6.2.3 Temperature profile across the Atlantic Ocean from the coast of Florida to the coast of Africa (inset). There is rapid temperature change near the surface in the thermocline zone, but the deep water temperature is fairly stable (eWOCE, http://www.ewoce.org/gallery/eWOCE_Tables.html#Atlantic).

Mid-latitude temperate regions show greater seasonal fluctuations in surface temperature than the poles or the tropics; an 8-15° C difference from summer to winter in temperate zones, compared to only ~2° C in polar and tropical areas. In temperate regions, the surface water is much warmer in the summer and the thermocline is more pronounced compared to the winter months. But in the winter the thermocline is *deeper* at mid-latitudes than it is in the summer. This is because winter storms churn up the surface water more than occurs in the summer, creating a deeper mixed layer and thus a deeper thermocline (Figure 6.2.5).

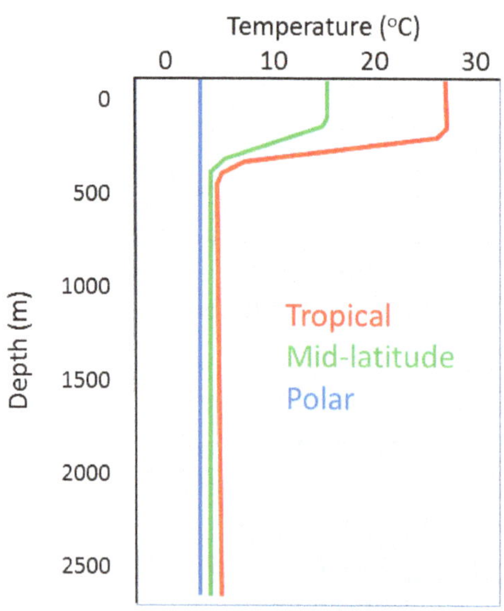

Figure 6.2.4 Representative temperature profiles for tropical, mid-latitude, and polar regions (PW).

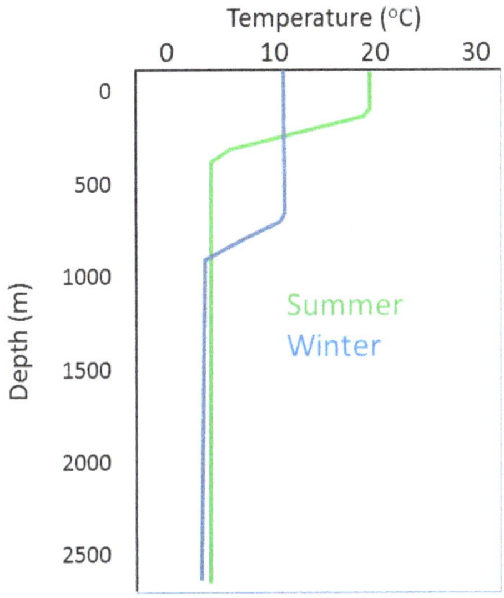

Figure 6.2.5 In temperate regions, the mixed layer is deeper and the thermocline less pronounced in the winter compared to the summer (PW).

Due to the high heat capacity of water, daily fluctuations in ocean temperature are fairly insignificant.

6.3 DENSITY

Density is the measure of mass per unit volume, typically expressed in grams per cubic centimeter (g/cm3). The density of fresh water is 1 g/cm3 at 4o C (see to section 5.1). However, the presence of salts and other dissolved substances raises the density of surface saltwater to a range of 1.02 to 1.03 g/cm3. Seawater's density can be heightened by lowering its temperature, elevating its salinity, or raising the pressure it is subjected to. The impact of pressure on density is minimal due to the relatively low compressibility of water. Therefore, pressure effects are only important at extreme depths. Without the small compression of water caused by pressure, the current sea level would be around 50 meters higher. The main elements that determine density are temperature and salinity, with temperature having the most significant influence. (Figure 6.3.1).

Since temperature has the greatest effect on density, density profiles are usually mirror images of temperature profiles (Figure 6.3.2). Density is lowest at the surface, where the water is the warmest. As depth increases, there is a region of rapidly increasing density with increasing depth, which is called the **pycnocline**. The pycnocline coincides with the thermocline, as it is the sudden decrease in temperature that leads to the increase in density. Below the pycnocline, density may be fairly constant (as is temperature), or it may continue to increase slightly towards the bottom.

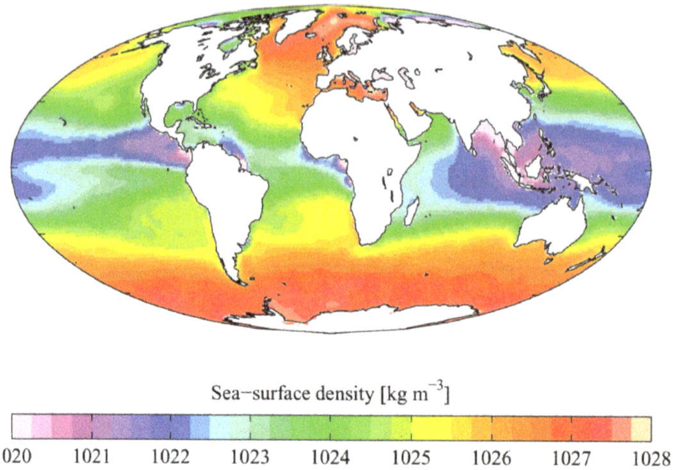

Figure 6.3.1 Global sea surface density. Colder polar regions display higher densities than warmer tropical zones (By Plumbago (Own work) [CC BY-SA 3.0, or GFDL (http://www.gnu.org/copyleft/fdl.html)], via Wikimedia Commons).

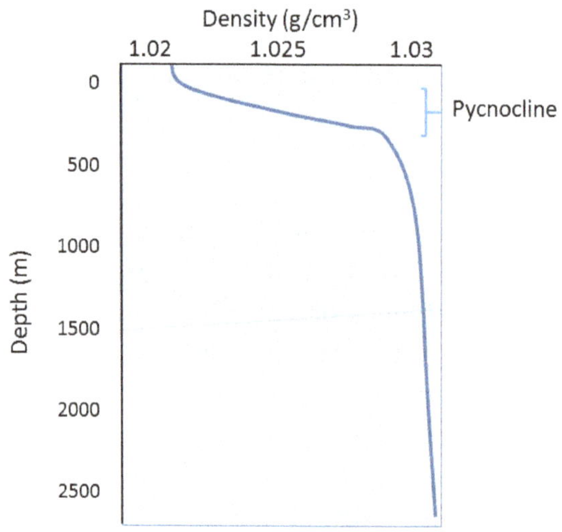

Figure 6.3.2 Representative density profile for the open ocean at mid-latitudes. The warm surface water causes a decrease in surface density (PW).

The profile depicted above illustrates a state of equilibrium, characterized by a significant level of differentiation, in which the warm, less dense layer is positioned above the colder, more dense layer. If denser water were to accumulate at the surface, the water masses would become unstable, causing the denser water to drop to the bottom and be replaced by less dense water at the surface. The vertical displacement of water masses due to variations in density, dictated by temperature and salinity, is known as thermohaline circulation. This concept is discussed in section 9.8. The thermocline and pycnocline work together to form a stratified water column, which acts as a barrier preventing the mixing of warmer, less dense surface water with colder, denser bottom water. Thus, the ascent of nutrient-rich deep water to the surface, which is crucial for sustaining primary production, may be hindered.

Similar to temperature, there are variations in density based on latitude. In tropical regions, the surface water is characterized by high temperatures and low density. It is distinctly separated from the colder and denser subsurface water by a well-defined thermocline. As mentioned before, this stratification hinders the upward movement of water rich in nutrients, leading to low productivity in tropical locations. In regions with high latitudes, the water maintains a consistently low temperature throughout its depths, resulting in minimal variation in density stratification. In polar locations, the absence of a pycnocline (or thermocline) facilitates the efficient mixing of cold, nutrient-rich deep water with the surface water, resulting in increased primary production.

6.4 SOUND

Sound is a form of energy transmitted through pressure waves; longitudinal or compressional waves similar to the seismic P-waves we discussed in section 3.3. With ocean sounds, the energy is transmitted via water molecules vibrating back and forth parallel to the direction of the sound wave, and passing on the energy to adjacent molecules. Therefore, sound travels faster and more efficiently when the molecules are closer together and are better able to transfer their energy to neighboring particles. In other words, sound travels faster through denser materials. Since water is much denser than air, the speed of sound in water (about 1500 m/s) is approximately five times faster than the speed in air (around 330 m/s). This helps explain why we sometimes have difficulty localizing the source of a sound that we hear underwater. We localize sound sources when our brains detect the tiny differences in the time of arrival of sounds reaching our ears. A sound coming from our left will reach our left ear a fraction of a second before reaching our right ear. Our brains can process that small difference in time of arrival to recognize the direction from which the sound came. In water, the sound is so much faster that the difference in arrival time between our ears becomes too small for us to interpret, and we lose the ability to localize the source.

However, as with sound in air, the speed of sound in the ocean is not constant; it is influenced by a number of variables including temperature, salinity, and pressure, and an increase in any of these factors will lead to an increase in the speed of sound. We have seen that these variables change with depth and location; so to will the speed of sound differ in different regions of the ocean.

To examine the way the speed of sound changes as a function of depth, we need to consider the vertical profiles for temperature and pressure. At the surface, the pressure is low, but the temperature is at its highest point in the water column. The temperature effects dominate at the surface, so the speed of sound is fast in surface waters. As depth increases, the temperature and the speed of sound decline. Near the bottom, the extreme pressure dominates, and even though temperatures are low, the speed of sound increases with depth. At moderate depths (between a few hundred and one thousand meters) there is a zone where both temperature and pressure are relatively low, so the speed of sound is at a minimum. This zone of minimum speed is called the **SOFAR channel** (**So**und **F**ixing **A**nd **R**anging) or the Deep Sound Channel (Figure 6.4.1).

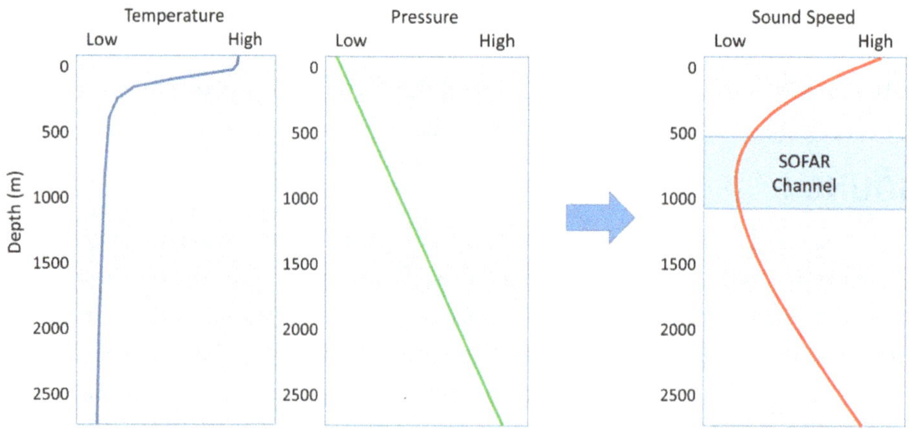

Figure 6.4.1 Profiles of temperature, pressure, and sound speed with depth. Sound speed is high at the surface due to the high temperatures, and is high at depth because of the high pressure. At moderate depths lies the SOFAR channel, the region of slowest sound speed (PW).

The significance of the SOFAR channel lies in its ability to facilitate the transmission of sounds over vast distances with minimal energy loss. Sound waves emitted within the channel propagate in a radial manner. When waves go from deeper to shallower water or vice versa outside of the sound channel, they are entering an area where sound travels quicker. As observed in the case of seismic waves, when sound waves come over an area where the speed of transmission varies, the waves are inclined to be refracted or bent back towards the area with a lower speed. Consequently, sound waves transitioning from the SOFAR channel to less deep water will experience refraction and be redirected back towards the channel. As the sound waves penetrate further down the channel, they will undergo refraction, causing them to bend upwards and return to the channel and the area where the speed of sound is slower. By confining the sound within the channel, a significant portion of it is prevented from dispersing in various directions into the water. Consequently, the sound can propagate over extensive distances with minimal energy loss. (Figure 6.4.2).

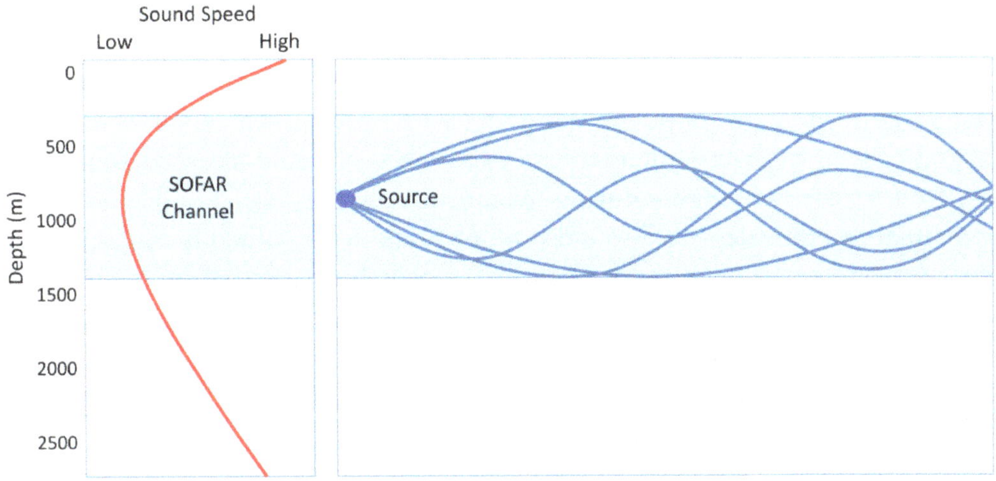

Figure 6.4.2 Sound propagation in the SOFAR channel. Sound waves emanating from a source will be refracted towards the region of slower speed, "trapping" the sound in the SOFAR channel (PW).

The SOFAR channel has numerous practical applications. Baleen whales are believed to utilize the SOFAR channel for long-range communication spanning hundreds to thousands of kilometers. Their vocalizations are characterized by their high volume and low frequency, enabling them to propagate across long distances in the water. The military has successfully employed the SOFAR channel to monitor submarines, and during World War II, it was utilized to identify pilots who had crashed or ships and planes that had gone missing. In the event of a pilot being stranded, they could deploy a compact apparatus into the sea. Once this device descends within the SOFAR channel, it would detonate, generating a sound that would be detectable at various listening stations. Triangulation can be used to determine the position of the source by analyzing the time of arrival of the sound at different receivers. In the 1990s, there was a proposal to utilize the SOFAR channel for the purpose of monitoring worldwide ocean temperatures. The ATOC (Acoustic Thermometry of Ocean Climate) project was intended to transmit loud, low frequency noises near Hawaii and California across the SOFAR channel to receiving sites around the Pacific. Scientists could track global changes in ocean temperatures by measuring the time it took for sounds to reach receivers. This is because sounds travel faster across a hotter ocean.

Due to the superior transmission properties of water compared to air, the energy needed to propagate a certain sound wave is greater in air than in water. The magnitude of sound, or its loudness, is quantified using the decibel (dB) scale. Transmitting sound through air requires approximately 61 times more energy than transmitting sound through water. Due to the disparity in energy levels, the transmission of sound through air and water results in a significant discrepancy of 61 dB. Consequently, a sound intensity of 120 dB in water would correspond to an approximate intensity of 60 dB in air. It is important to consider this when attempting to compare sounds in the ocean with noises in the air. A sound level of 130 dB in air is about similar to the experience of being situated 100 meters away from an aircraft engine during takeoff. A sound level of 130 dB in water is approximately equivalent to a sound level of 70 dB in air, which is comparable to the strength of a vacuum cleaner's sound. It is worth noting that on the dB scale, a 10 dB rise corresponds to a sound that is 10 times louder. To clarify, a sound that measures 20 dB is ten times more intense than a sound that measures 10 dB, and a sound that is 30 dB is one hundred times more intense than a sound that measures 10 dB.

6.5 LIGHT

Radiant energy from the sun is important for several major oceanic processes:

- Climate, winds, and major ocean currents are ultimately dependent on solar radiation reaching the Earth and heating different areas to different degrees.
- Sunlight warms the surface water where much oceanic life lives.
- Solar radiation provides light for photosynthesis, which supports the entire ocean ecosystem.

The energy reaching Earth from the sun is a form of electromagnetic radiation, which is represented by the electromagnetic spectrum (Figure 6.5.1). Electromagnetic waves vary in their frequency and wavelength. High frequency waves have very short wavelengths, and are very high energy forms of radiation, such as gamma rays and x-rays. These rays can easily penetrate the bodies of living organisms and interfere with individual atoms and molecules. At the other end of the spectrum are low energy, long wavelength waves such as radio waves, which do not pose a hazard to living organisms.

Most of the solar energy reaching the Earth is in the range of visible light, with wavelengths between about 400-700 nm. Each color of visible light has a unique wavelength, and together they make up white light. The shortest wavelengths are on the violet and ultraviolet end of the spectrum, while the longest wavelengths are at the red and infrared end. In between, the colors of the visible spectrum comprise the familiar "ROYGBIV"; red, orange, yellow, green, blue, indigo, and violet.

Water is very effective at absorbing incoming light, so the amount of light penetrating the ocean declines rapidly (is attenuated) with depth (Figure 6.5.2). At 1 m depth, only 45% of the solar energy that falls on the ocean surface remains. At

10 m depth only 16% of the light is still present, and only 1% of the original light is left at 100 m. No light penetrates beyond 1000 m.

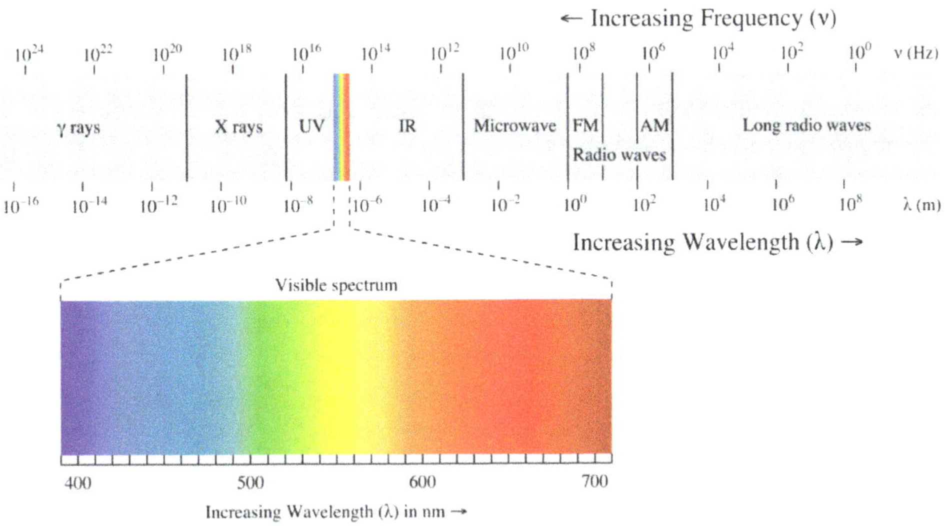

Figure 6.5.1 The electromagnetic spectrum. Frequency is expressed in Hertz (Hz), or waves per second, while wavelengths are expressed in meters (Phillip Roman, CC BY-SA 3.0, via Wikimedia Commons).

In addition to overall attenuation, the oceans absorb the different wavelengths of light at different rates (Figure 6.5.2). The wavelengths at the extreme ends of the visible spectrum are attenuated faster than those wavelengths in the middle. Longer wavelengths are absorbed first; red is absorbed in the upper 10 m, orange by about 40 m, and yellow disappears before 100 m. Shorter wavelengths penetrate further, with blue and green light reaching the deepest depths.

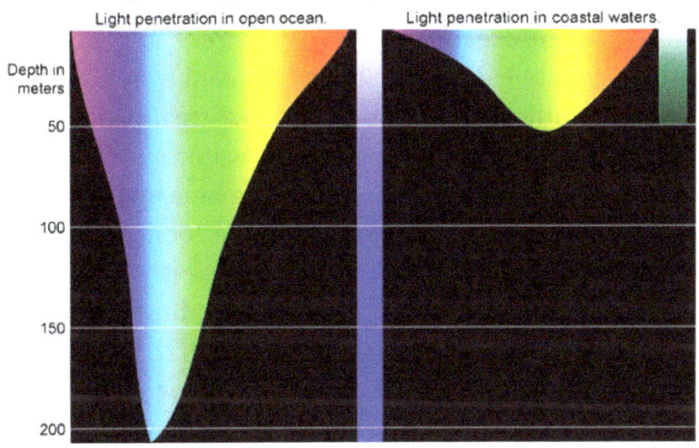

Figure 6.5.2 Light penetration in open ocean and coastal water, showing the different depths to which each color will penetrate (By NOAA – National Oceanic and Atmospheric Administration [Public domain], via Wikimedia Commons).

This explains why everything appears blue under water. The colors we perceive depends on the wavelengths of light that are received by our eyes. If an object appears red to us, that is because the object reflects red light but absorbs all of the other colors. So the only color reaching our eyes is red. Under water, blue is the only color of light still available at depth, so that is the only color that can be reflected back to our eyes, and everything has a blue tinge under water. A red object at depth will not appear red to us because there is no red light available to reflect off of the object. Objects in water will only appear as their real colors near the surface where all wavelengths of light are still available, or if the other wavelengths of light are provided artificially, such as by illuminating the object with a dive light.

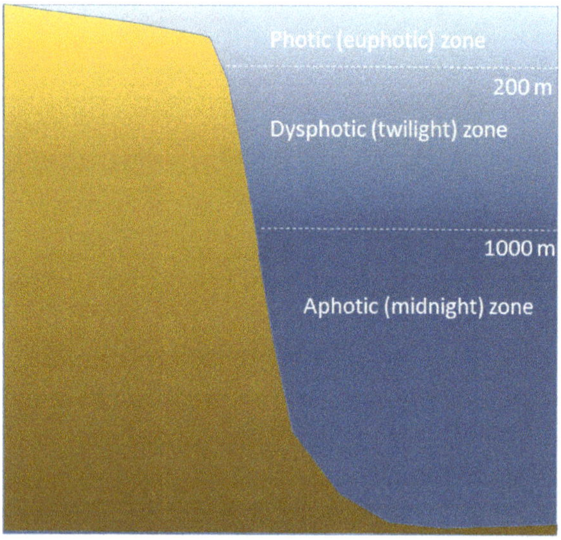

Figure 6.5.3 The zones of the water column as defined by the amount of light penetration (PW).

The water in the open ocean appears transparent and azure due to its low concentration of particulate matter, such as phytoplankton or suspended particles. The clarity of the water allows light to penetrate to greater depths. The blue light is able to penetrate deeply and is dispersed by the water molecules, whilst all other hues are absorbed, resulting in the water appearing blue. In contrast, coastal water frequently exhibits a greenish hue (Figure 6.5.2). The concentration of suspended sediment, algae, and microscopic creatures is significantly higher in coastal water compared to the open ocean. Several species, including phytoplankton, utilize their photosynthetic pigments to absorb light mostly in the blue and red spectrum, resulting in green light being the most prominent wavelength of reflected light. Consequently, when the proportion of phytoplankton in water is higher, the water appears greener. In the presence of significant quantities of suspended particles, small silt particles can also absorb blue light, resulting in a further deviation of the water hue from blue.

The ocean can be categorized into several depth layers based on the extent of light penetration, as explained in section 1.3 (Figure 6.5.3). The top 200 meters is known as the photic or euphotic zone. The region referred to here is the area where there is sufficient light for photosynthesis to occur. This region is known as the epipelagic zone. The dysphotic zone, often known as the twilight zone, is located between depths of 200 to 1000 meters, which corresponds to the mesopelagic zone. At these depths, there is insufficient light to sustain photosynthesis. The aphotic (or midnight) zone is located below 1000 m and is characterized by the absence of light penetration. This area encompasses the vast majority of the ocean's volume and is characterized by a complete absence of light.

Chapter 7 PRIMARY PRODUCTION

7.1 PRIMARY PRODUCTION

Primary production refers to the process of generating fresh organic matter from inorganic substances. This organic matter forms the foundation of the food chain for the majority of marine organisms. Primary production typically pertains to the photosynthesis process, where plants and algae utilize light energy to make chemical fuels, as described by the following reaction:

$$6CO_2 + 6H_2O \rightarrow C_6H_{12}O_6 + 6O_2$$

In this process, facilitated by light energy, carbon dioxide and water react to form glucose and oxygen. Nevertheless, bacteria are capable of conducting primary generation by chemosynthesis even in the absence of light. Instead of relying on light as an energy source for the reaction, the energy is derived from the oxidation of inorganic substances, such as hydrogen sulfide. (see section 4.11 on hydrothermal vents). Our focus will be on photosynthesis as it significantly contributes to the overall productivity of the ocean compared to chemosynthesis.

The organisms responsible for oceanic primary production include a wide diversity of marine plants and algae. While many people may be more familiar with the larger seagrasses and macroalgae (seaweeds), by far the greatest amount of photosynthesis in the ocean comes from microscopic algae, the **phytoplankton**. The term "plankton" refers to organisms that drift with the currents, and the phytoplankton are the free-floating algae that undergo photosynthesis (contrast this with the **zooplankton**, who are the drifting animals). In section 7.2 we will take a closer look at the organisms responsible for oceanic primary production.

The aggregate quantity of organic matter generated by the producers is referred to as the gross primary productivity, or total production. Nevertheless, the primary producers utilize a portion of this organic matter for their own respiration, resulting in the remaining amount available to sustain the consumers in the ocean, known as net production (gross productivity minus respiration). The gross production can be categorized into two distinct components: fresh production and regenerated output. External nutrients, such as those transported through upwelling or ocean currents, contribute to the growth of new organisms in the local environment. The recycling of nutrients within an ecosystem leads to the regeneration of production outputs.

In general, the amount of biological activity in the ocean is comparable to that on land. The annual net production of marine organisms is approximately 35-50 billion metric tons, whereas terrestrial organisms produce around 50-70 billion tons per year. Nevertheless, the amount of biomass accountable for this production in the ocean is only 1-2 billion metric tons, in contrast to the 600-1000 billion metric tons of biomass found in terrestrial systems. The oceans are generating about the same quantity of organic material as terrestrial producers, but having significantly less producer biomass. One factor contributing to this difference is the continuous consumption of phytoplankton, whereas a significant portion of terrestrial biomass has a longer lifespan compared to plankton.

7.2 THE PRODUCERS

Although the phytoplankton are microscopic in size compared to marine plants and macroalgae like seaweeds and seagrasses, they account for by far the greatest amount of photosynthesis in the oceans; about 95% of all marine primary productivity. Most of the production by phytoplankton comes from three groups, the diatoms, dinoflagellates, and coccolithophores.

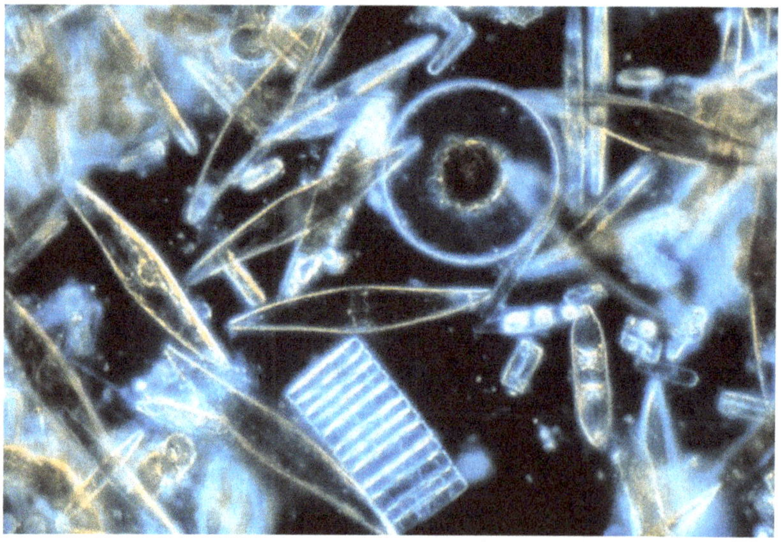

Figure 7.2.1 A variety of diatom species from the Southern Ocean, seen under a light microscope (Prof. Gordon T. Taylor, Stony Brook University, [Public domain], via Wikimedia Commons).

Diatoms are single-celled algae consisting of cellular material inside a shell, or test, made of silica, a component of glass. Diatoms are relatively large, reaching up to about 1 mm in diameter, and come in a range of shapes, from circular disks to elongated or triangular forms (Figure 7.2.1). In some species of diatoms individual cells link together into multicellular chains. Diatoms are very efficient producers, with up to 55% of the absorbed sunlight energy incorporated into carbohydrate formation; this is one of the highest photosynthetic efficiencies known. Diatoms are most abundant in coastal and cold, nutrient-rich waters. Where diatoms are bountiful, the underlying sediments are rich in their silica shells, creating siliceous sediment and diatomaceous earth (see section 12.3).

Dinoflagellates are another form of single-shelled photosynthetic algae that are generally smaller than diatoms, most with sizes in the 0.015 - 0.04 mm range. Most dinoflagellates have a characteristic pair of flagella (hence the name); small whip-like "tails" that they use for locomotion. Usually there is one flagellum that trails from the body to provide forward movement, and another that encircles the cell to make it spin as it moves. Unlike diatoms, dinoflagellates do not have a mineralized shell. Instead, many are covered by cellulose, which easily decomposes in seawater, so their shells do not really contribute to sediment formation (Figure 7.2.2). While most dinoflagellates undergo photosynthesis, some species will also ingest prey.

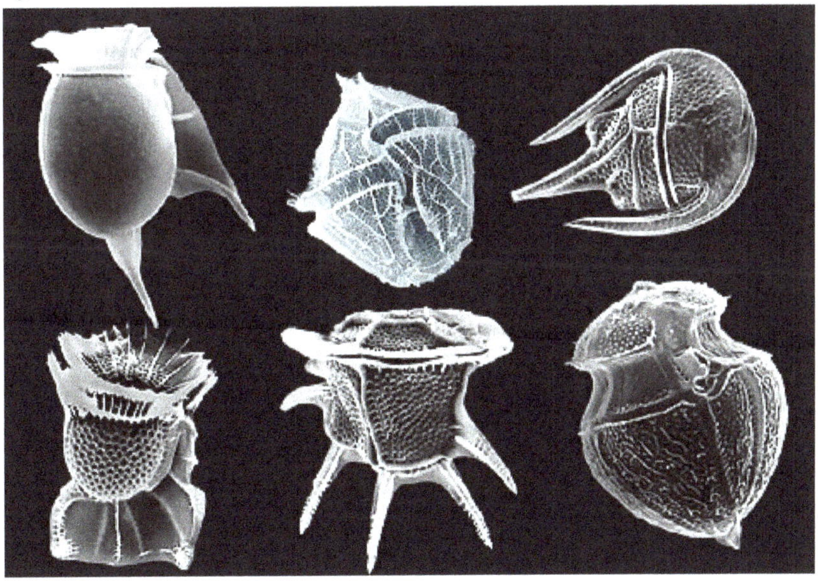

Figure 7.2.2 Electron microscope image of several dinoflagellate species. In some of the species, a groove can be seen running around the circumference of the organism; this groove is usually home to one of the flagella (fickleandfreckled, https://www.flickr.com/photos/fickleandfreckled/6939384773, CC-BY 2.0).

A third, much smaller type of phytoplankton includes the **coccolithophores**, which range from about 5-100 micrometers wide. As with diatoms and dinoflagellates, these are single-celled photosynthetic algae that contribute significantly to oceanic primary production, but their cellular material is encased in a very different kind of shell. The test (shell) is made up of a number of interlocking circular plates composed of calcium carbonate that link together to form a sphere (Figure 7.2.3). Coccolithophores are most abundant in warm, open ocean waters, and their sinking tests can lead to calcium carbonate sediments in some parts of the ocean (see section 12.3).

Figure 7.2.3 Electron microscope image of the North Atlantic coccolithophore *Coccolithus pelagicus* (Richard Lampitt, Jeremy Young, The Natural History Museum, London (http://planktonnet.awi.de/) [CC BY 2.5], via Wikimedia Commons).

Recent evidence suggests that another group of organisms, the **bacteria**, or **picoplankton**, may be very important primary producers. Although they are very small, on the order of 0.2-2 micrometers long, they can be found in very high concentrations, and may be responsible for up to 70% of all productivity in some parts of the ocean.

Harmful algal blooms

Primary production provides plentiful food resources for ocean consumers, so a high abundance of phytoplankton is a good thing, right? As in many other cases, too much of a good thing can sometimes be dangerous, and an overabundance of dinoflagellates or diatoms can often create serious concerns. These events are referred to as **harmful algal blooms**, or HABs. HABs can occur for a number of reasons, although a common one is an overabundance of nutrients, which is often due to excessive terrestrial runoff of fertilizers or other nitrogen- and phosphate-containing materials. These conditions lead to an explosion in algal populations that can change the color of the water if the cells are in high enough concentrations. Figure 7.2.4 shows a massive bloom that contained so many dinoflagellate cells that it turned the water reddish-brown, a so-called "**red tide.**" (It has been suggested that Biblical references to seas being "turned to blood" may have actually been describing red tide events).

Figure 7.2.4 A red tide caused by dinoflagellates near the Scripps Institution of Oceanography Pier, La Jolla California (Alejandro Díaz, Public Domain via Wikimedia Commons).

These massive algal blooms can have some serious consequences. For one, when all of this algae eventually dies off and sinks, their decomposition uses up the dissolved oxygen in the water, leaving anoxic or hypoxic conditions that can lead to the mass die-off of fish and invertebrates. Dinoflagellates and diatoms are also capable of producing toxins themselves. These phytoplankton get eaten by fish, shellfish and other organisms, and in high abundances the toxins become concentrated in the tissues of the consumers. When humans or other higher-level consumers then eat these organisms, the toxins are concentrated enough to cause sickness or even death. For example, some dinoflagellates produce a toxin that causes **paralytic shellfish poisoning**, which can occur in humans as soon as 30 minutes after eating infected shellfish. This toxin attacks the nervous system, producing symptoms of dizziness, nausea, slurred speech, loss of feeling, and uncoordinated movements, and can ultimately be fatal. Diatoms produce a toxin called domoic acid that causes **amnesic shellfish poisoning**, leading to memory loss, seizures, and potentially death. Domoic acid poisoning also affects marine animals; it is thought to have been responsible for an event in Capitola, California in 1961 where flocks of seabirds acted crazily, even attacking humans. This event inspired the Alfred Hitchcock movie "The Birds."

7.3 FACTORS INFLUENCING PRODUCTION

Various elements influence the productivity of terrestrial plants, such as light, temperature, nutrients, soil composition, and water availability. Phytoplankton do not require soil and are not affected by water supply. The ocean normally maintains a more consistent temperature compared to land. Therefore, the productivity of phytoplankton mostly depends on the presence of light and nutrients.

Light

Since light is vital for photosynthesis, phytoplankton and other primary producers are limited to the uppermost layers of the ocean where light is abundant enough to sustain the reaction. As depth increases, light intensity decreases until there reaches a depth where photosynthesis can no longer occur (Figure 7.3.1). The region through which sufficient light for photosynthesis can penetrate is called the photic or euphotic zone, which extends down to about 200 m (section 6.5).

In addition to undergoing photosynthesis, phytoplankton also respire, consuming some of the organic compounds they produce. Rates of respiration are not light dependent, and respiration occurs at all depths and light levels. Therefore, as depth increases the rate of photosynthesis declines as light is diminished, until a point is reached where the rate of photosynthesis equals the respiration rate (Figure 7.3.1). This depth is the **compensation depth**, and it marks the lower level of the photic zone, and represents the depth where net primary production ends. Below this depth, there is net respiration.

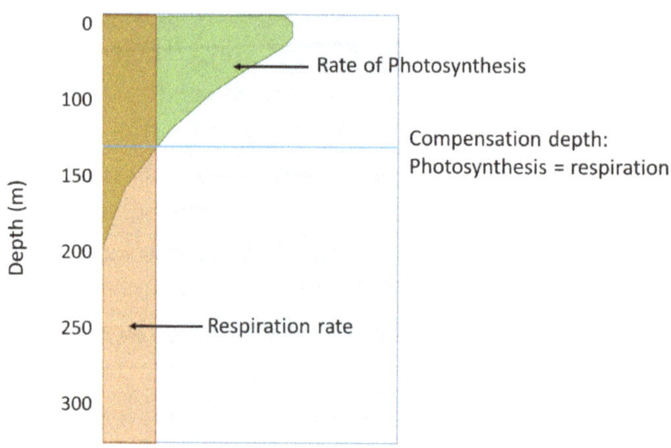

Figure 7.3.1 As depth increases the rate of photosynthesis declines as light becomes limited. The rate of respiration remains consistent at all depths. The depth where photosynthesis equals respiration is the compensation depth (PW).

Nutrients

Nutrients are required by all of the marine primary producers. The major nutrients required by phytoplankton are nitrogen and phosphorus, in the forms of nitrate

NO_3^-, nitrite NO_2^{2-}, ammonium NH_4^+, and phosphate PO_4^{3-}. Many phytoplankton, particularly the diatoms, also require silica, SiO_2, for shell formation. All of these nutrients occur in very small amounts in seawater, so they are often the limiting factors for phytoplankton growth in most situations, particularly the nitrogen compounds. For example, agricultural soil contains 0.5% nitrogen in the upper meter of soil, while surface ocean water contains about 0.00005% nitrogen, 1/10,000 the amount in soil.

As we saw in section 5.6, nutrients are not distributed evenly throughout the water column (Figure 7.3.2). Near the surface nutrients are quickly utilized by phytoplankton as they become available, so surface waters are nutrient-poor. But as the phytoplankton are consumed or die they are recycled into particles of organic matter, such as fecal pellets or carcasses, that sink into deeper water. Once in deep water, decomposition of these materials releases the nutrients back into the water column. Because there are no producers to utilize them at depth, nutrients are more abundant in deeper water.

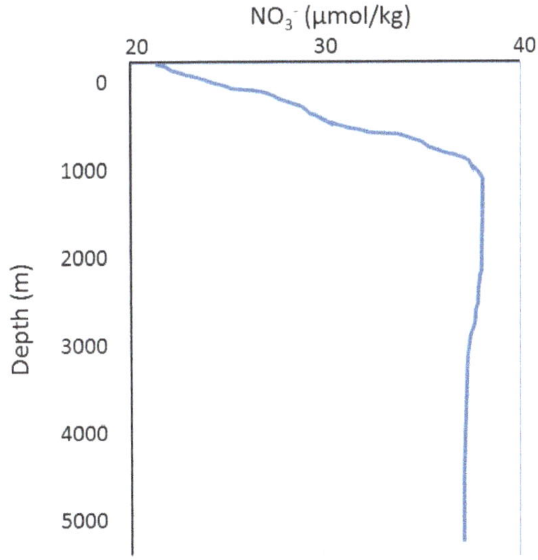

Figure 7.3.2 Representative nutrient profile in the open ocean. While this profile shows nitrate abundance, the profile is similar for other nutrients such as phosphate and silica (PW).

These deep water nutrients are out of reach of the phytoplankton at the surface. The thermocline and density stratification of the water column generally prevents the nutrient-rich deep water from mixing with the surface water. However, under certain conditions this nutrient-rich deep water may be brought to the surface through the process of upwelling (see section 9.5). Where upwelling occurs there is usually high productivity as the phytoplankton can take advantage of the input of nutrients.

8 Chapter | OCEANS AND CLIMATE

8.1 EARTH'S HEAT BUDGET

The equilibrium between the amount of heat entering and leaving Earth is known as its heat budget. In order to maintain thermal equilibrium, it is necessary to balance the budget by ensuring that the amount of heat entering a system is equal to the amount of heat leaving the system. Below is the thermal energy balance of Earth. (Figure 8.1.1).

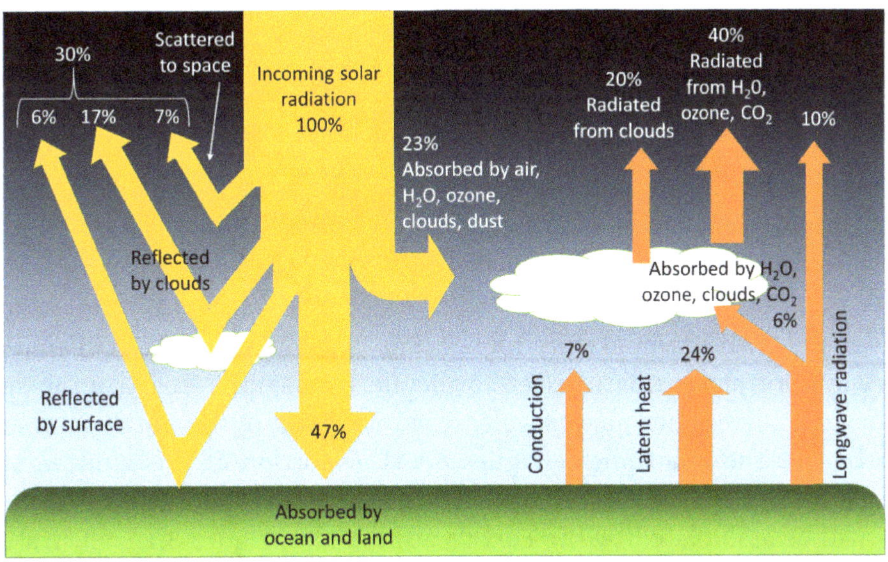

Figure 8.1.1 Earth's heat budget. Of all of the solar radiation reaching Earth, 30% is reflected back to space and 70% is absorbed by the Earth (47%) and atmosphere (23%). The heat absorbed by the land and oceans is exchanged with the atmosphere through conduction, radiation, and latent heat (phase change). The heat absorbed by the atmosphere is eventually radiated back into space (PW).

Approximately 30% of the solar energy that reaches the Earth is redirected back into space by the atmosphere, clouds, and Earth's surface. An further 23%

of the energy is absorbed by the water vapor, clouds, and dust in the atmosphere, where it is transformed into heat. Approximately 47% of the solar radiation that reaches the Earth is absorbed by the land and ocean. This absorbed energy causes an increase in temperature on the Earth's surface. The Earth absorbs energy which is subsequently released back into the atmosphere by three distinct processes: conduction, radiation, and latent heat (phase change) (as shown in Figure 8.1.1).

Conduction refers to the process of heat transfer that occurs when there is direct contact between a surface and the surrounding atmosphere. Air has a low thermal conductivity, making it an effective insulator. Consequently, conduction plays a little role in the transfer of energy between the Earth and the atmosphere, accounting for around 7% of the incoming solar energy.

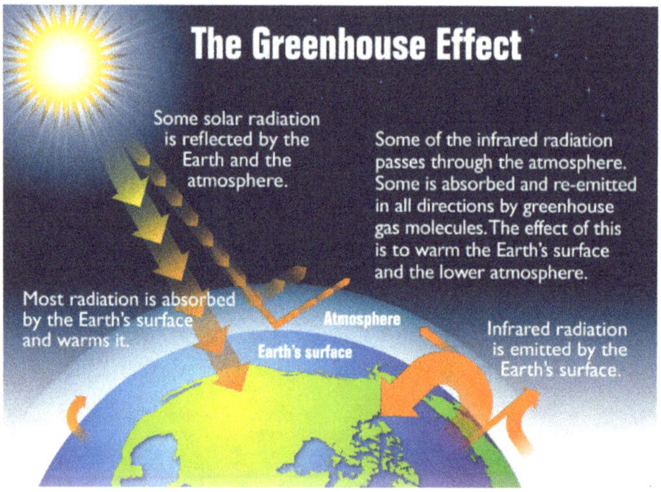

Figure 8.1.2 An explanation of the greenhouse effect (By US EPA [Public domain], via Wikimedia Commons).

Any object with a temperature higher than absolute zero (-273o C) emits heat as longwave, infrared radiation (see to the electromagnetic spectrum in section 6.5). The Earth, like other objects, releases around 16% of the initial solar energy it absorbs into the atmosphere (Figure 8.1.1). A portion of this emitted energy will disperse into the vacuum of space, while a substantial quantity of thermal energy will be assimilated by the Earth's atmosphere. Figure 8.1.2 illustrates the fundamental principle behind the greenhouse effect. During the greenhouse effect, the Earth's surface absorbs shortwave solar radiation that has passed through the atmosphere. After being emitted by the Earth, the radiation transforms into long wavelength, infrared radiation, which encounters difficulty in passing through the atmosphere. However, the atmosphere absorbs this infrared radiation, specifically through the greenhouse gases like CO_2, methane, and water vapor. Consequently, the atmosphere becomes warmer. In the absence of the greenhouse effect, the Earth's average temperature would plummet to approximately -18°C, rendering it too frigid to sustain liquid water and consequently making life as we currently understand it impossible.

There is widespread apprehension regarding the greenhouse effect worldwide, not due to its mere existence, but due to its escalating intensity, leading to climate change or global warming. Since the onset of the Industrial Revolution, there has been a significant rise in the atmospheric concentrations of important greenhouse gases, notably CO_2 and methane. This increase can be attributed to industrialization, the combustion of fossil fuels, and deforestation. Simultaneously, there has been a swift increase in the worldwide climate's temperature; carbon dioxide concentrations have surged by over 25% and the global temperature has escalated by 0.5 degrees Celsius in the last century. If the production of these greenhouse gases is not reduced, the current rapid warming trend may persist, leading to potentially severe consequences. Refer to section 8.5 for comprehensive information regarding the causes and consequences of climate change.

The primary mechanism for transferring heat between the land or oceans and the atmosphere is through latent heat, which is released or absorbed during phase changes of water, such as when it transitions between solid, liquid, and vapor states (see to section 5.1). Adding heat to liquid water causes it to undergo evaporation. As water vapor is produced, the heat is extracted from the ocean and transferred to the atmosphere along with the water vapor. When water vapor undergoes condensation and transforms into rain, the heat is then sent back to the oceans. The identical process occurs with the production and dissolution of ice. The process of melting ice absorbs heat, whereas the formation of ice releases heat. These phase shifts facilitate the flow of heat between the oceans and the atmosphere.

In order to maintain the thermal balance, the atmosphere absorbs heat from solar radiation and through processes such as conduction, radiation, and latent heat. Eventually, this absorbed heat is sent back into space (Figure 8.1.1).

Differential Heating of Earth's Surface

If the Earth were a flat plane directly exposed to the sun, every region of the plane would get an equal amount of solar energy. Due to the Earth's spherical shape, sunlight is unevenly dispersed throughout its surface, resulting in varying degrees of heating in different locations. The differential heating of Earth's surface happens due to many factors. Due to the Earth's curvature, sunlight only strikes the surface perpendicularly at the center of the sphere, specifically in the equatorial regions. At any other location on Earth, the angle between the surface and the incoming solar radiation is always less than 90 degrees. As a result, the equator will experience a higher concentration of incoming solar radiation within a smaller region, whereas the poles will have the same amount of radiation spread out over a broader area

(Figure 8.1.3). Therefore, the tropical regions experience higher levels of solar radiation and a greater amount of heat per unit of surface area compared to the polar regions.

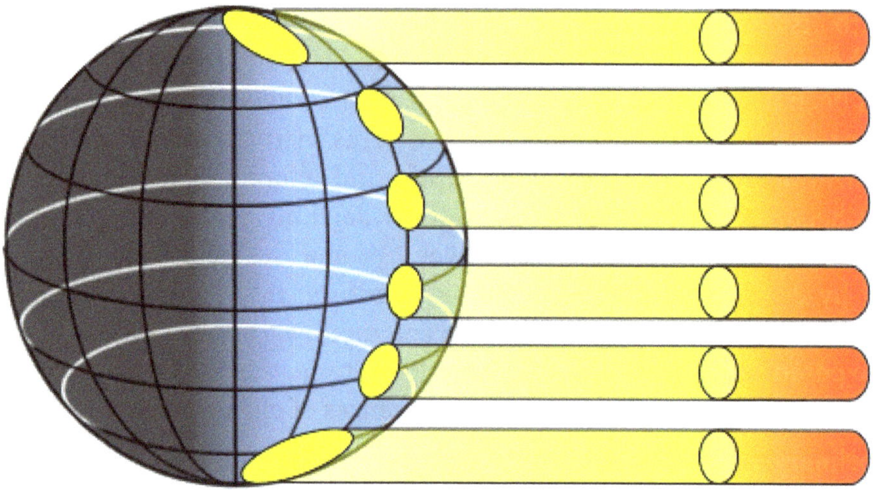

Figure 8.1.3 Because of the curvature of the Earth, the same amount of sunlight will be spread out over a larger area at the poles compared to the equator. The equator therefore receives more intense sunlight, and a greater amount of heat per unit of area (By Thebiologyprimer (Own work) [CC0], via Wikimedia Commons).

The angle of incidence of sunlight on the Earth's surface plays a role in causing uneven heating in an extra manner. At the poles, due to the oblique angle of solar radiation, a greater proportion of light will be deflected off the surface and atmosphere, resulting in its reflection back into space. At the equator, the incidence angle of light is such that a greater proportion of the energy is absorbed rather than reflected by the surface. The poles have a higher albedo, which means they reflect more solar energy compared to other portions of the Earth. Albedo is the measure of how much light a surface reflects. Lighter surfaces exhibit greater reflectivity compared to darker surfaces, which have a higher capacity for energy absorption, resulting in a higher albedo. In polar regions, the presence of ice, snow, and cloud cover results in a significantly higher albedo. As a result, the poles reflect a greater amount of solar energy and absorb less compared to lower latitudes. Due to these many causes, the poles receive a significantly lower amount of solar radiation compared to the equatorial regions. This is the reason why the poles experience chilly temperatures while the tropics are characterized by high levels of warmth.

However, there is a fascinating nuance to this worldwide dispersion of heat. Tropical regions get a greater amount of radiant heat than they emit, while the poles emit more heat than they receive (Figure 8.1.4). Consequently, it is reasonable to anticipate a persistent increase in temperature in the tropics, while the poles experience a growing coldness. However, this is not the situation; thus, what is occurring? Instead of the heat staying confined to the equator, around 20% of the heat from the tropics gets transferred to the poles before it is released. The extensive transportation of energy helps to regulate the climates in both extreme conditions. The processes involved in the movement of heat include ocean and atmospheric circulation, which will be discussed in the following section.

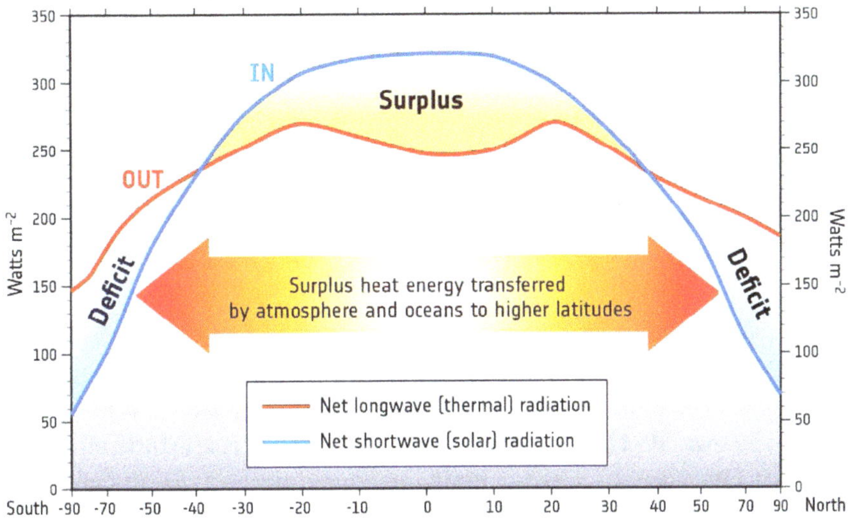

Figure 8.1.4 The balance between heat gain and heat loss as a function of latitude. Excess heat received near the equator is transferred towards the poles (National Oceanography Centre (NOC). Creative Commons 3.0 unported license).

The idea of differential heating of the Earth's surface is fundamental to understanding a wide range of oceanographic and atmospheric processes. This differential heating leads to atmospheric convection, which creates winds, which blow over the water and create waves and surface currents, and these currents influence nutrient distribution, which promotes primary production, which then supports the rest of the ocean ecosystem. So there's a lot riding on the simple fact that more light reaches the tropics than the poles!

8.2 WINDS AND THE CORIOLIS EFFECT

Differential heating of the Earth's surface results in equatorial regions receiving more heat than the poles (section 8.1). As air is warmed at the equator it becomes less dense and rises, while at the poles the cold air is denser and sinks. If the Earth was non-rotating, the warm air rising at the equator would reach the upper atmosphere and begin moving horizontally towards the poles. As the air reached the poles it would cool and sink, and would move over the surface of Earth back towards the equator. This would result in one large atmospheric convection cell in each hemisphere (Figure 8.2.1), with air rising at the equator and sinking at the poles, and the movement of air over the Earth's surface creating the winds. On this non-rotating Earth, the prevailing winds would thus blow from the poles towards the equator in both hemispheres (Figure 8.2.1).

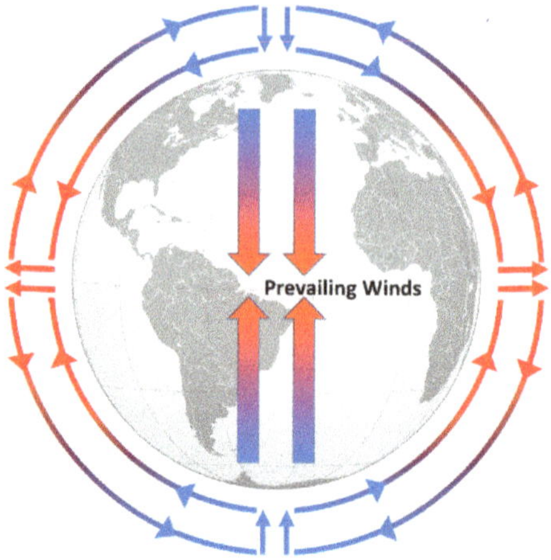

Figure 8.2.1 Hypothetical atmospheric convection cells on a non-rotating Earth. Air rises at the equator and sinks at the poles, creating a single convection cell in each hemisphere. The prevailing winds moving over the Earth's surface blow from the poles towards the equator in both hemispheres (Modified by PW from globe image by Location_of_Cape_Verde_in_the_globe.svg: Eddo derivative work: Luan fala! [CC BY-SA 3.0], via Wikimedia Commons).

The non-rotating situation in Figure 8.2.1 is of course only hypothetical, and in reality the Earth's rotation makes this atmospheric circulation a bit more complex. The paths of the winds on a rotating Earth are deflected by the **Coriolis Effect.** The Coriolis Effect is a result of the fact that different latitudes on Earth rotate at different speeds. This is because every point on Earth must make a complete rotation in 24 hours, but some points must travel farther, and therefore faster, to complete the rotation in the same amount of time. In 24 hours a point on the equator must complete a rotation distance equal to the circumference of the Earth, which is about 40,000 km. A point right on the poles covers no distance in that time; it just turns in a circle. So the speed of rotation at the equator is about 1600 km/hr, while at the poles the speed is 0 km/hr. Latitudes in between rotate at intermediate speeds; approximately 1400 km/hr at 30° and 800 km/hr at 60°. As objects move over the surface of the Earth they encounter regions of varying speed, which causes their path to be deflected by the Coriolis Effect.

To explain the Coriolis Effect, imagine a cannon positioned at the equator and facing north. Even though the cannon appears stationary to someone on Earth, it is in fact moving east at about 1600 km/hr due to Earth's rotation. When the cannon fires the projectile travels north towards its target; but it also continues to move to the east at 1600 km/hr, the speed it had while it was still in the cannon. As the shell moves over higher latitudes, its momentum carries it eastward faster than the speed at which the ground beneath it is rotating. For example, by 30° latitude the shell is moving east at 1600 km/hr while the ground is moving east at only 1400 km/hr. Therefore, the shell gets "ahead" of its target, and will land to the east of its intended destination. From the point of view of the cannon, the path of the projectile appears to have been deflected to the right (red arrow, Figure 8.2.2). Similarly, a cannon located at 60° and facing the equator will be moving east at 800 km/hr. When its shell is fired towards the equator, the shell will be moving east at 800 km/hr, but as it approaches the equator it will be moving over land that is traveling east *faster* than the projectile. So the projectile gets "behind" its target, and will land to the west of its destination. But from the point of view of the cannon facing the equator, the path of the shell still appears to have been deflected to the right (green arrow, Figure 8.2.2). Therefore, in the Northern Hemisphere, the apparent Coriolis deflection will *always* be to the **right**.

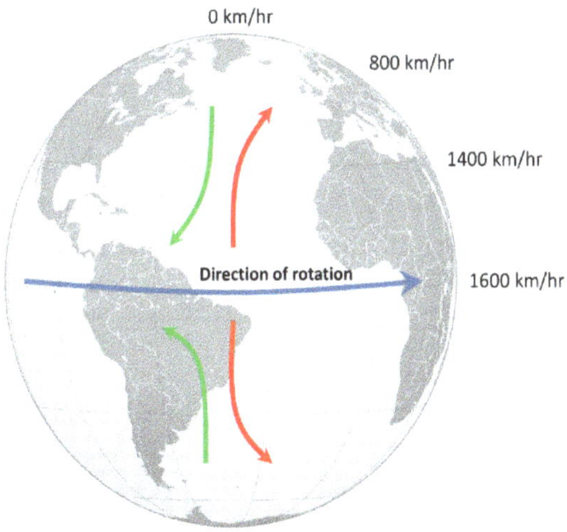

Figure 8.2.2 The Coriolis Effect. Objects moving from the equator towards the poles (red arrows) move into a region of slower rotational speed and their paths are deflected "ahead" of their point of origin. Movement from high latitudes to low latitudes (green arrows) goes from a region of low speed to a region of higher rotation speed, and there is deflection "behind" their point of origin. In the Northern Hemisphere this deflection is always to the right from the point of origin, and in the Southern Hemisphere the deflection is always to the left (Modified by PW from globe image by Location_of_Cape_Verde_in_the_globe.svg: Eddo derivative work: Luan fala! [CC BY-SA 3.0], via Wikimedia Commons).

In the Southern Hemisphere the situation is reversed (Figure 8.2.2). Objects moving towards the equator from the south pole are moving from low speed to high speed, so are left behind and their path is deflected to the left. Movement from the equator towards the south pole also leads to deflection to the left. In the Southern Hemisphere, the Coriolis deflection is always to the **left** from the point of origin.

The magnitude of the Coriolis deflection is related to the difference in rotation speed between the start and end points. Between the poles and 60° latitude, the difference in rotation speed is 800 km/hr. Between the equator and 30° latitude, the difference is only 200 km/hr (Figure 8.2.2). Therefore the strength of the Coriolis Effect is stronger near the poles, and weaker at the equator.

Because of the rotation of the Earth and the Coriolis Effect, rather than a single atmospheric convection cell in each hemisphere, there are three major cells per hemisphere. Warm air rising at the equator cools as it moves through the upper atmosphere, and it descends at around 30° latitude. The convection cells created by rising air at the equator and sinking air at 30° are referred to as **Hadley Cells**, of which there is one in each hemisphere. The cold air that descends at the poles moves over the Earth's surface towards the equator, and by about 60° latitude it begins to rise, creating a **Polar Cell** between 60° and 90°. Between 30° and 60° lie the **Ferrel Cells**, composed of sinking air at 30° and rising air at 60° (Figure 8.2.3). With three convection cells in each hemisphere that rotate in alternate directions, the surface winds no longer always blow from the poles towards the equator as in the non-rotating Earth in Figure 8.2.1. Instead, surface winds in both hemispheres blow towards the equator between 90° and 60° latitude, and between 0° and 30° latitude. Between 30° and 60° latitude, the surface winds blow towards the poles (Figure 8.2.3).

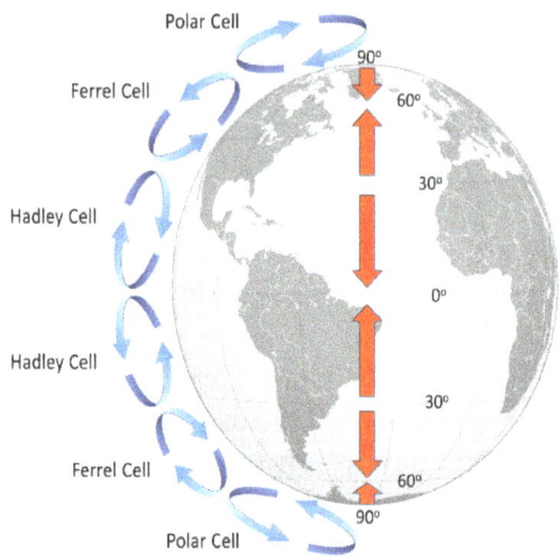

Figure 8.2.3 On a rotating Earth, there are three atmospheric convection cells in each hemisphere, leading to alternating bands of surface winds (red arrows) (Modified by PW from globe image by Location_of_Cape_Verde_in_the_globe.svg: Eddo derivative work: Luan fala! [CC BY-SA 3.0], via Wikimedia Commons).

The surface winds created by the atmospheric convection cells are also influenced by the Coriolis Effect as they change latitudes. The Coriolis Effect deflects the path of the winds to the right in the Northern Hemisphere and to the left in the Southern Hemisphere. Adding this deflection leads to the pattern of prevailing winds illustrated in Figure 8.2.4. Between the equator and 30° latitude are the **trade winds**; the northeast trade winds in the Northern Hemisphere and the southeast trade winds in the Southern Hemisphere (note that winds are named based on the direction from which they originate, not where they are going). The **westerlies** are the dominant winds between 30° and 60° in both hemispheres, and the **polar easterlies** are found between 60° and the poles.

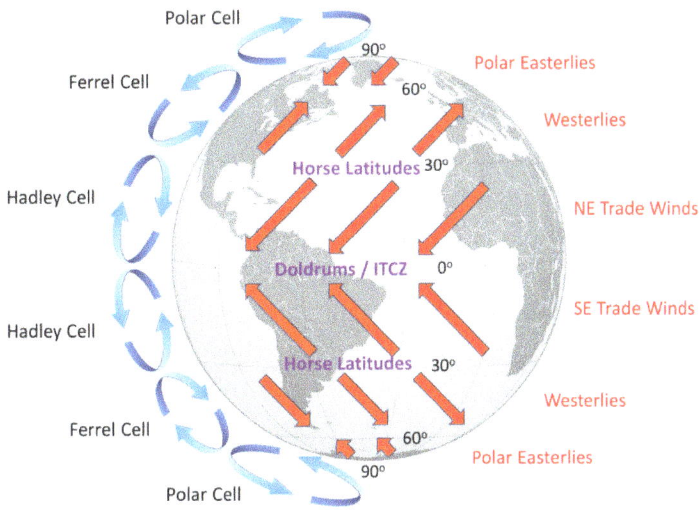

Figure 8.2.4 The prevailing wind patterns of Earth (Modified by PW from globe image by Location_of_Cape_Verde_in_the_globe.svg: Eddo derivative work: Luan fala! [CC BY-SA 3.0], via Wikimedia Commons).

In between these wind bands lie regions of high and low pressure. High pressure zones occur where air is descending, while low pressure zones indicate rising air. Along the equator the rising air creates a low pressure region called the **doldrums**, or the **Intertropical Convergence Zone** (ITCZ)(convergence zone because this is where the trade winds converge). At 30° latitude there are high pressure zones of descending air known as the **horse latitudes**, or the subtropical highs. Finally, at 60° lies another low pressure region called the **polar front.** It should be noted that these high and low pressure zones are not fixed in place; their latitude fluctuates depending on the season, and these fluctuations have important implications for regional climates.

Doldrums? Horse latitudes? Trade winds?

These may seem like some odd names for these atmospheric phenomena, but many of them can be traced back to maritime traditions and lore.

The **doldrums** refer to regions of low pressure around the equator. In these areas, air is rising rather than moving horizontally, so these regions commonly encounter very light winds. The lack of wind could leave sailing ships becalmed for days or weeks at a time, which was not good for the morale of the ship's crew.

Like the doldrums the **horse latitudes** are also areas with light winds, this time due to descending air, which could leave ships becalmed. One explanation for the term "horse latitudes" is that when these ships became stranded they ran the risk of running out of food or water. To conserve these resources, sailors would throw their dead or dying horses overboard, hence the "horse latitudes." Another explanation is that many sailors received part of their pay before a voyage, and often spent it before departing. This meant that they would spend the first part of the voyage working without pay and in debt, a period called the "dead horse" time, which might last for a few months. When they started earning their pay once again, they had a "dead horse" ceremony and threw a pretend horse overboard. The timing of this ceremony often coincided with reaching the horse latitudes, leading to the association of the ceremony with the location. A third explanation is that a ship was referred to as "horsed" when winds were weak and the ship instead had to rely on ocean currents to move them. This could be a common occurrence in the high pressure zones around 30° latitude, so they were referred to as the horse latitudes.

The term **trade winds** may have originally derived from the terms for "track" or "path", but the term may have become more common during European exploration and commercialization of the New World. Mariners sailing from Europe to the New World could sail south until they reached the trade winds, which would then propel their ships across the Atlantic to the Caribbean. To return to Europe, ships could sail to the northeast until they entered the westerlies, which would then steer them back to Europe.

8.3 WINDS AND CLIMATE

In the previous section we learned that rising air creates low pressure systems, and sinking air creates high pressure. In addition to their role in creating the surface winds, these high and low pressure systems also influence other climatic phenomena. Along the equator air is rising as it is warmed by solar radiation (section 8.2). Warm air contains more water vapor than cold air, which is why we experience humidity during the summer and not during the winter. The water content of air roughly doubles with every 10° C increase in temperature. So the air rising at the equator is warm and full of water vapor; as it rises into the upper atmosphere it cools, and the cool air can no longer hold as much water vapor, so the water condenses and forms rain. Therefore, low pressure systems are associated with precipitation, and we see wet habitats like tropical rainforests near the equator (Figure 8.3.1).

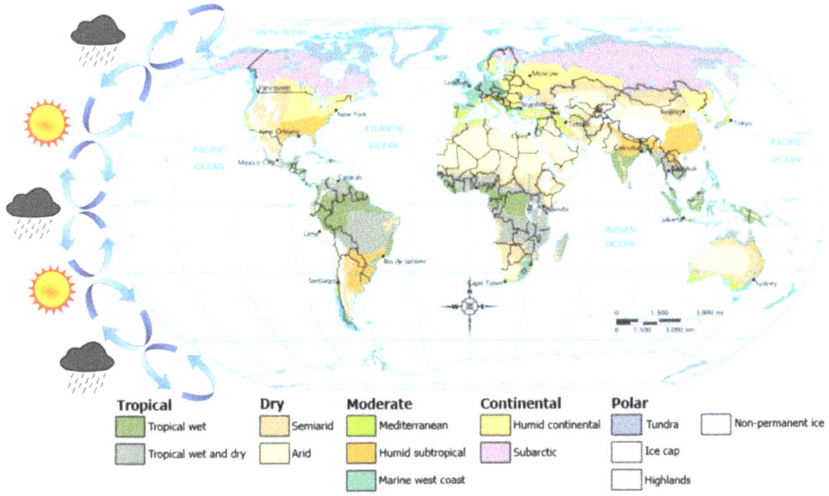

Figure 8.3.1 Major global climatic regions in relation to atmospheric convection cells. Rising air and low pressure creates rain and wet environments at 0° and 60° latitudes, while high pressure, sinking air creates drier conditions at 30° and 90° latitudes. (Modified by PW; Map by Waitak at en.wikipedia Later version(s) were uploaded by Splette at en.wikipedia; Sun by Inductiveload (Own work Based on File:Nuvola_apps_kweather.svg); Raincloud by Calusarul (Own work); all [CC BY-SA 3.0], via Wikimedia Commons).

After rising and producing rain near the equator, the air masses move towards 30° latitude and sink back towards Earth as part of the Hadley convection cells. This air has lost most of its moisture after producing the equatorial rains, so the sinking air is dry, resulting in arid climates near 30° latitude in both hemispheres. Many of the major desert regions on Earth are located near 30° latitude, including much of Australia, the Middle East, and the Sahara Desert of Africa (Figure 8.3.1). The air also becomes compressed and heats up as it sinks, absorbing any moisture from the clouds and creating clear skies. Thus high pressure systems are associated with dry weather and clear skies. This cycle of high and low pressure regions continues with the Ferrel and Polar convection cells, leading to rain and the boreal forests at 60° latitude in the Northern Hemisphere (there are no corresponding large land masses at these latitudes in the Southern Hemisphere). At the poles, descending, dry air produces little precipitation, leading to the polar desert climate.

The elevation of the land also plays a role in precipitation and climactic characteristics. As moist air moves over land and encounters mountains it rises, expands, and cools because of the declining pressure and temperature. The cool air holds less water vapor, so condensation occurs and rain falls on the windward side of the mountains. As the air passes over the mountains to the leeward side, it is now dry air, and as it sinks the pressure increases, it heats back up, any moisture revaporizes, and it creates dry, deserts regions behind the mountains (Figure 8.3.2).

This phenomenon is referred to as a **rain shadow**, and can be found in areas such as the Tibetan Plateau and Gobi Desert behind the Himalayas, Death Valley behind the Sierra Nevada mountains, and the dry San Joaquin Valley in California.

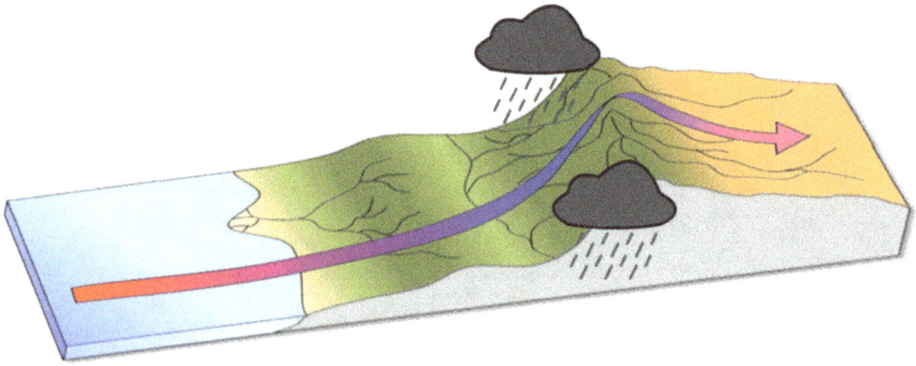

Figure 8.3.2 A rain shadow. Air rising over mountains cools and condenses and forms rain, leaving dry descending air and arid conditions on the other side of the mountain (Modified by PW from Thebiologyprimer, Public domain via Wikimedia Commons).

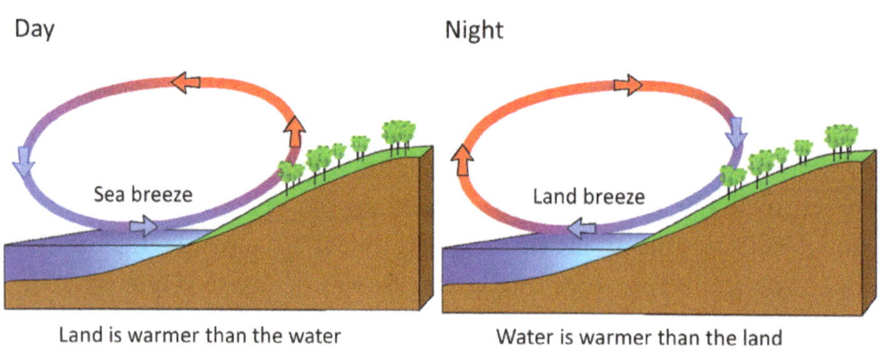

Figure 8.3.3 Land heats and cools faster than the ocean, so during the day the land is warmer than the water leading to rising air over land and a sea breeze. At night, the ocean is warmer than the land, creating a land breeze (Modified by PW derivative work: Ingwik (Diagrama de formacion de la brisa-breeze.png) [CC-BY-SA-3.0 or GFDL (http://www.gnu.org/copyleft/fdl.html)], via Wikimedia Commons).

Rising and falling air are also responsible for more localized, short-term wind patterns in coastal areas. Due to the high heat capacity of water, land heats up and cools down about five times faster than water. During the day the sun heats up the land faster than it heats the water, setting up a convection cell of warmer rising air over the land and sinking cooler air over the water. This creates winds blowing from the water towards the land during the day and early evening; a **sea breeze** (Figure 8.3.3). The opposite occurs at night, when the land cools more quickly than the ocean. Now the ocean is warmer than the land, so air rises over the water and sinks over the land, creating a convection cell where winds blow from land towards the water. This is a **land breeze**, which blows at night and into the early morning (Figure 8.3.3).

The same phenomenon leads to seasonal climatic changes in many areas. During the winter the lower pressure is over the warmer ocean, and the high pressure is over the colder land, so winds blow from land to sea. In summer the land is warmer than the ocean, causing low pressure over the land and winds to blow from the ocean towards the land. The winds blowing from the ocean contain a lot of water vapor, and as the moist air passes over land and rises, it cools and condenses causing seasonal rains, such as the summer **monsoons** of southeast Asia (Figure 8.3.4).

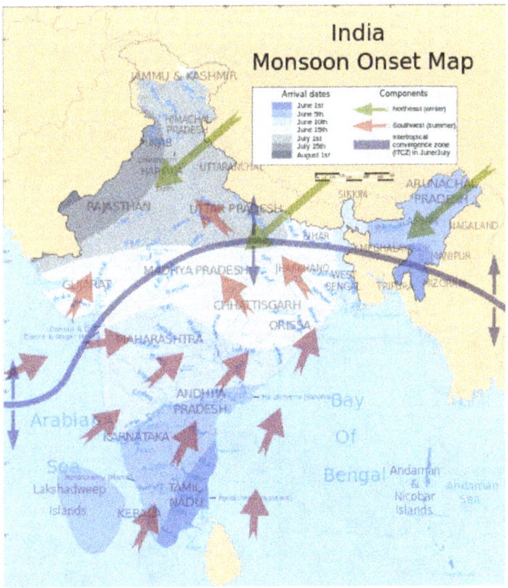

Figure 8.3.4 Seasonal wind patterns and monsoons over India. In summer, moist air from the ocean moves over the continent and rises, creating rain and the summer monsoons (pink arrows). In winter, winds are blowing from land to the sea, leading to the dry season (green arrows) (By Saravask, based on work by Planemad and Nichalp [CC BY-SA 3.0], via Wikimedia Commons).

8.4 HURRICANES

The most dramatic examples of low pressure systems leading to storms and rain are hurricanes, cyclones, and typhoons. All three of these terms describe the same atmospheric processes and the same types of storms; it's just that different terminology is used in different parts of the world. In the Atlantic and Northeast Pacific, the storms are called hurricanes, in the Indian and South Pacific Oceans they are referred to as cyclones, and they are called typhoons in the Northwest Pacific.

Hurricanes begin as low pressure systems formed over warm, tropical water. They only form in tropical regions because they need the heat from the warm water to fuel the storm. The warm, moist air rises, cools, and condenses, forming rain, and the condensation releases more latent heat into the atmosphere. This heat causes even more air to rise and condense, further fueling the storm.

As the air rises towards the center of the storm, more warm tropical air rushes in to replace it, causing very strong winds. But the air does not move directly towards the center of the storm. Because of the large size of hurricanes, the air rushing towards the center will be deflected by the Coriolis Effect, causing the entire storm to rotate. In the Northern Hemisphere that deflection is to the right, causing Northern Hemisphere hurricanes to rotate counterclockwise. In the Southern Hemisphere, the winds are deflected to the left, leading to a clockwise rotation (Figure 8.4.1).

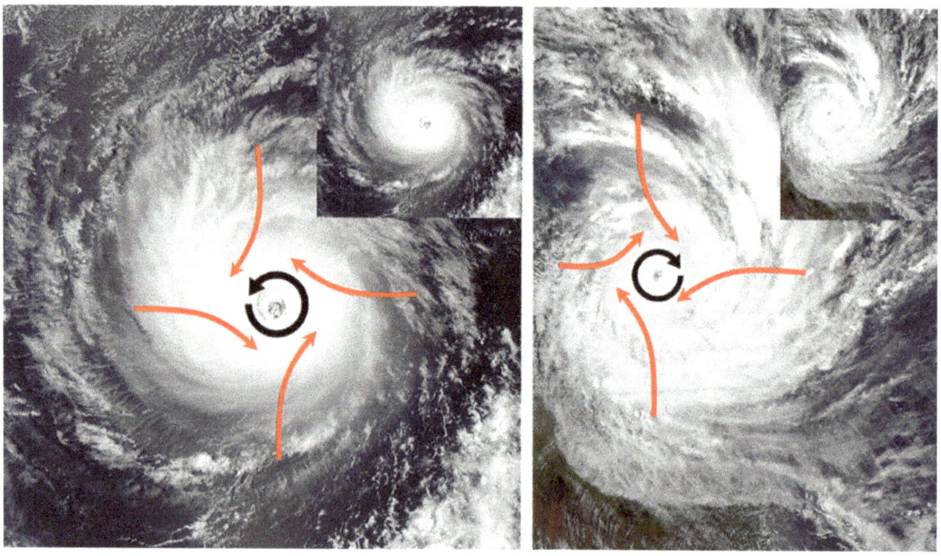

Figure 8.4.1 Hurricanes in the Northern Hemisphere rotate counterclockwise (left, Hurricane Daniel, 2006), as air rushes towards the center and is deflected to the right by the Coriolis Effect. In the Southern Hemisphere, hurricanes rotate clockwise as the Coriolis deflection is to the left (right, Cyclone Yasi, 2011) (Modified by PW; Daniel image by NASA image courtesy Jeff Schmaltz, MODIS Land Rapid Response Team at NASA GSFC; Yasi image by NASA; MODIS [Public domain], via Wikimedia Commons).

The violent winds characteristic of hurricanes are the result of the spiraling air that is moving towards the center of the storm, and once its winds exceed 74 mph the storm officially becomes a hurricane. At the very center of the hurricane, the pressure is so low that cool, dry air from the upper atmosphere get sucked downwards, leading to a central region of calm, clear skies; the hurricane's eye (Figure 8.4.2).

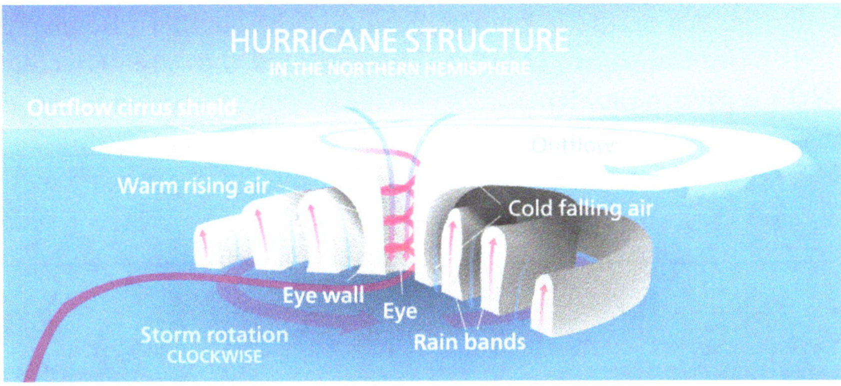

Figure 8.4.2 Hurricane structure. Air rising in the center of the hurricane is replaced by warm air moving inwards, and the Coriolis Effect deflects the winds, causing the storm to rotate. In the eye, the extreme low pressure causes cool, dry air to sink, creating calm, clear conditions within the eye (By Kelvinsong (Own work) [CC BY 3.0], via Wikimedia Commons).

Hurricanes in the North Atlantic form as tropical storms over the warm water off of the African coast, and are moved east to west by the trade winds (Figure 8.4.3). As the storms move west over the tropical ocean, their energy increases until they reach hurricane status. As they approach the Caribbean, the Coriolis Effect deflects their path to the right, causing them to move towards the north (Figure 8.4.3). Eventually hurricanes might make landfall, causing extensive damage to coastal areas through the high winds, rain, and flooding. However, hurricanes often die out fairly soon after reaching land. When a storm moves over land it becomes cut off from the warm moist ocean air that has sustained it. Without that fuel source, the storm loses power and begins to dissipate.

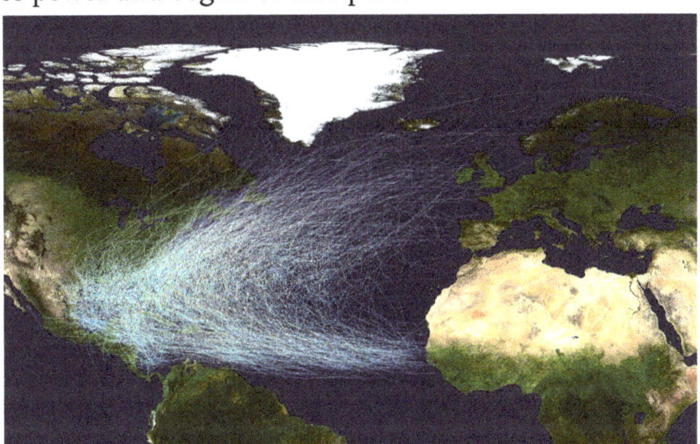

Figure 8.4.3 Hurricane tracks in the North Atlantic from 1980-2005. Hurricanes begin near the coast of Africa and are blown westward by the trade winds. Coriolis deflection causes them to take a northward path as they approach the Caribbean (By Nilfanion [Public domain], via Wikimedia Commons).

A similar pattern occurs in the Pacific and in the Southern Hemisphere. The trade winds move the storms from east to west, and they are deflected as they approach the coasts; to the right in the Northern Hemisphere and to the left in the Southern Hemisphere (Figure 8.4.4).

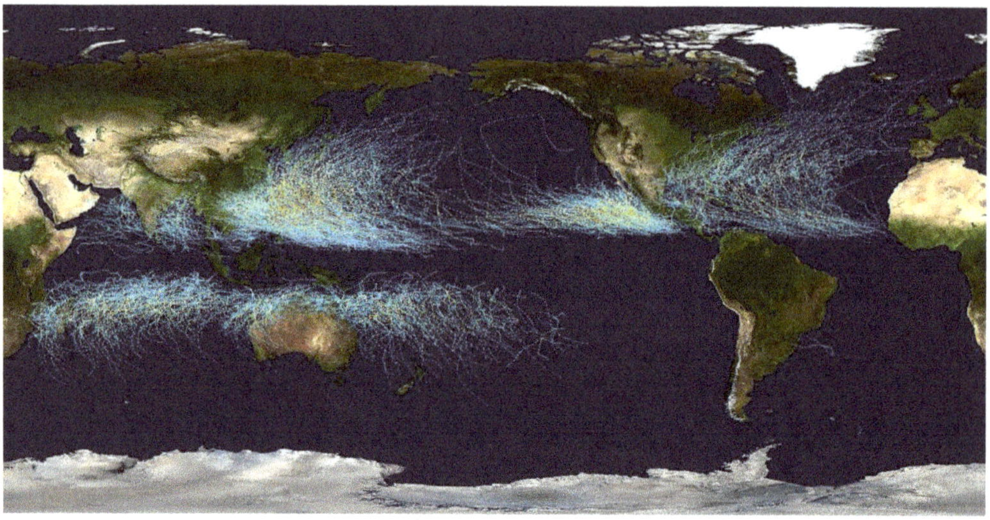

Figure 8.4.4 Global hurricane/cyclone tracks 1985-2005. Hurricanes move west via the trade winds, and their paths are deflected away from the equator in both hemispheres as they approach the continents (Background image: NASA this version: Nilfanion [Public domain], via Wikimedia Commons).

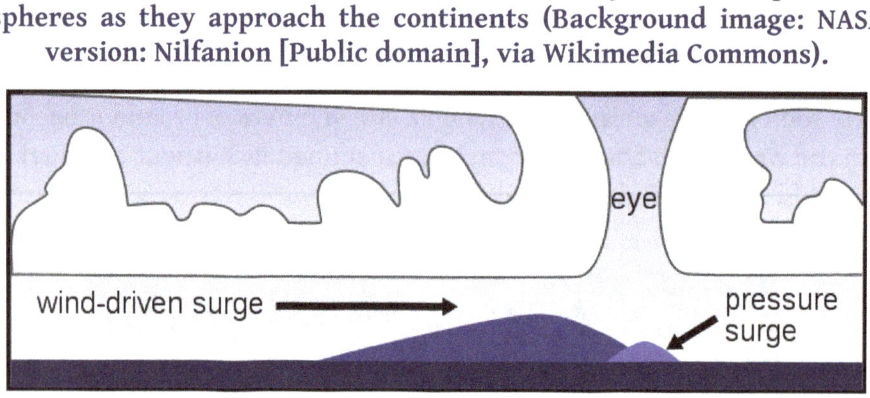

Figure 8.4.5 Storm surges are created by hurricanes and move with the storm, causing a rapid rise in sea level when they reach shore. Pressure surges are due to the low pressure within the hurricane's eye, wind-driven surges are a product of high winds piling up water (By Howcheng. Original graphic by Robert Simmon, NASA GSFC. [Public domain], via Wikimedia Commons).

While the very high winds and intense rain of hurricanes can cause significant damage, in many cases it is the **storm surge** that leads to the most death and destruction. The storm surge is a "hill" of water that forms on the ocean surface below a hurricane. The surge is the result of two processes; a small hill is produced due to the extreme low pressure in the eye of a hurricane, which pulls water upwards towards the eye, creating a pressure surge. A larger surge is produced by the winds blowing and piling up water in the direction the storm is traveling (Figure 8.4.5). As the hurricane makes landfall, the effect of the storm surge is equivalent to a very large and sudden rise in sea level as the surge moves over the land, causing extensive flooding.

In 1970 the Bhola Cyclone struck Bangladesh with a 40 ft. storm surge, leading to the death of about 500,000 people, the deadliest hurricane in history. The east coast of the United States was hit by the New England Hurricane of 1938, which had a 16 ft. storm surge and left almost 700 people dead.

Preventing Storm Surge Damage

In response to the hurricane-related tragedies like those listed above, may cities have built hurricane barriers designed to reduce the flooding and damage associated with storm surges. Downtown Providence, Rhode Island, USA, was submerged under 13 feet of water during the Great New England hurricane of 1938, and was flooded again following Hurricane Carol in 1954. In the 1960s the Fox Point Hurricane Barrier was constructed at the mouth of the Providence River. It consists of a high wall with three "doors" that are left open under normal conditions, but can be closed during a hurricane to prevent a storm surge of up to 20.5 feet from inundating the city (Figure 8.4.6, left). A related concept is seen in the storm surge barrier on the Hollandse IJssel river in the Netherlands, where the barrier is lowered to prevent flooding (Figure 8.4.6, right).

Figure 8.4.6 The Fox Point Hurricane Barrier in Providence, RI, USA (left) and the surge barrier on the Hollandse IJssel river in southern Holland. (Fox Point by Marcbela (Marc N. Belanger) (Own work) [Public domain], via Wikimedia Commons; Dutch barrier by Mark Voorendt (Own work) [CC BY-SA 4.0], via Wikimedia Commons).

8.5 CLIMATE CHANGE

*Modified from "Physical Geology" by Steven Earle**

If one thing has been constant about Earth's climate over geological time, it is its constant change. In the geological record, we can see this in the evidence of glaciations in the distant past, and we can also detect periods of extreme warmth by looking at the isotope composition of seafloor sediments. Not only has the climate changed frequently, the temperature fluctuations have been very significant. Today's mean global temperature is about 15°C. However, during its coldest periods, the global mean was as cold as -50°C, while at various times during the Paleozoic and Mesozoic and during the Paleocene–Eocene thermal maximum, it was close to 30°C.

There are two parts to climate change, the first one is known as **climate forcing**, which is when conditions change to give the climate a little nudge in one direction or the other. The second part of climate change, and the one that typically does most of the work, is what we call a **feedback**. When a climate forcing changes the climate a little, a whole series of environmental changes take place, many of which either exaggerate the initial change (**positive feedback**), or suppress the change (**negative feedback**).

An example of a climate forcing mechanism is the increase in the amount of carbon dioxide (CO_2) in the atmosphere that results from our use of fossil fuels. CO_2 traps heat in the atmosphere and leads to climate warming. Warming changes vegetation patterns; contributes to the melting of snow, ice, and permafrost; causes sea level to rise; reduces the solubility of CO_2 in sea water; and has a number of other minor effects. Most of these changes contribute to more warming. Melting of permafrost, for example, is a strong positive feedback because frozen soil contains trapped organic matter that is converted to CO_2 and methane (CH_4) when the soil thaws. Both these gases accumulate in the atmosphere and add to the warming effect. On the other hand, if warming causes more vegetation growth, that vegetation should absorb CO_2, thus reducing the warming effect, which would be a negative feedback. Under our current conditions — a planet that still has lots of glacial ice and permafrost — most of the feedbacks that result from a warming climate are positive feedbacks and so the climate changes that we cause get naturally amplified by natural processes.

Natural Climate Forcing

Natural climate forcing has been going on throughout geological time. A wide range of processes has been operating at widely different time scales, from a few years to billions of years. The longest-term natural forcing variation is related to the evolution of the Sun. Like most other stars of a similar mass, our Sun is evolving. For the past 4.6 billion years, its rate of nuclear fusion has been increasing, and it is now emitting about 40% more energy (as light) than it did at the beginning of geological time. A difference of 40% is big, so it's a little surprising that the temperature on Earth has remained at a reasonable and habitable temperature for all of this time. The mechanism for that relative climate stability has been the evolution of our atmosphere from one that was dominated by CO_2, and also had significant levels of CH_4 — both greenhouse gasses — to one with only a few hundred parts per million of CO_2 and just under 1 part per million of CH_4. Those changes to our atmosphere have been no accident; over geological time, life and its metabolic processes have evolved (such as the evolution of photosynthetic bacteria that consume CO_2) and changed the atmosphere to conditions that remained cool enough to be habitable.

The position of the Earth relative to the Sun is another important component of natural climate forcing. Earth's orbit around the Sun is nearly circular, but like all physical systems, it has natural oscillations. First, the shape of the orbit changes on a regular time scale (close to 100,000 years) from being close to circular to being very slightly elliptical. But the circularity of the orbit is not what matters; it is the fact that as the orbit becomes more elliptical, the position of the Sun within that ellipse becomes less central or more **eccentric** (Figure 8.5.1a). **Eccentricity** is important because when it is high, the Earth-Sun distance varies more from season to season than it does when eccentricity is low.

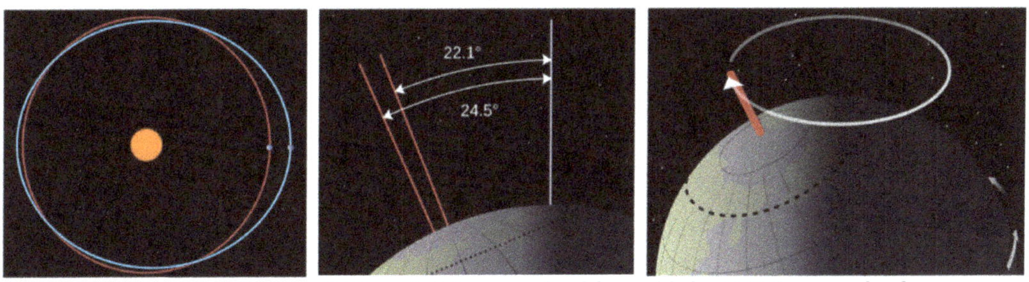

(a) The 100,000 year cycle of eccentricity of the Sun within the Earth's orbit. The blue orbit is more elliptical than the red one, and the Sun is offset from the centre of the ellipse

(b) The 41,000 year cycle of the angle of tilt (obliquity) of the Earth's axis (red lines) compared to a line perpendicular to the plane of the Earth's orbit (light blue line).

(c) the 20,000 year cycle of precession of the Earth's rotational axis

Figure 8.5.1 Components of the Milankovitch cycles, which influence global climate over thousands of years (Steven Earle, "Physical Geology").

Second, Earth rotates around an axis through the North and South Poles, and that axis is at an angle to the plane of Earth's orbit around the Sun (Figure 8.5.1b). The angle of tilt (also known as **obliquity**) varies on a time scale of 41,000 years. When the angle is at its maximum (24.5°), Earth's seasonal differences are accentuated. When the angle is at its minimum (22.1°), seasonal differences are minimized. The current hypothesis is that glaciation is favored at low seasonal differences as summers would be cooler and snow would be less likely to melt and more likely to accumulate from year to year. Third, the direction in which Earth's rotational axis points also varies, on a time scale of about 20,000 years (Figure 8.5.1c). This variation, known as **precession**, means that although the North Pole is presently pointing to the star Polaris (the pole star), in 10,000 years it will point to the star Vega. The importance of eccentricity, tilt, and precession to Earth's climate cycles (now known as **Milankovitch Cycles**) was first pointed out by Yugoslavian engineer and mathematician Milutin Milankovitch in the early 1900s. Milankovitch recognized that although the variations in the orbital cycles did not affect the total amount of insolation (light energy from the Sun) that Earth received, it did affect where on Earth that energy was strongest.

Volcanic eruptions don't just involve lava flows and exploding rock fragments; various particulates and gases are also released, the important ones being sulphur dioxide and CO_2. Sulphur dioxide is an aerosol that reflects incoming solar radiation and has a net cooling effect that is short lived (a few years in most cases, as the particulates settle out of the atmosphere within a couple of years), and doesn't typically contribute to longer-term climate change. Volcanic CO_2 emissions can contribute to climate warming but only if a greater-than-average level of volcanism is sustained over a long time (at least tens of thousands of years). It is widely believed that the catastrophic end-Permian extinction (at 250 Ma) resulted from warming initiated by the eruption of the massive Siberian Traps over a period of at least a million years.

Ocean currents are important to climate, and currents also have a tendency to oscillate. Glacial ice cores show clear evidence of changes in the Gulf Stream that affected global climate on a time scale of about 1,500 years during the last glaciation. The east-west changes in sea-surface temperature and surface pressure in the equatorial Pacific Ocean, known as the El Niño Southern Oscillation or ENSO (see section 9.6) varies on a much shorter time scale of between two and seven years. These variations tend to garner the attention of the public because they have significant climate implications in many parts of the world. The strongest El Niños in recent decades were in 1983, 1998, and 2015 and those were very warm years from a global perspective. During a strong El Niño, the equatorial Pacific sea-surface temperatures are warmer than normal and heat the atmosphere above the ocean, which leads to warmer-than-average global temperatures.

Climate Feedbacks

As already stated, climate feedbacks are critically important in amplifying weak climate forcings into full-blown climate changes. Since Earth still has a very large volume of ice, mostly in the continental ice sheets of Antarctica and Greenland, but also in alpine glaciers and permafrost, melting is one of the key feedback mechanisms. Melting of ice and snow leads to several different types of feedbacks, an important one being a change in albedo, or the reflectivity of a surface. Earth's various surfaces have widely differing albedos, expressed as the percentage of light that reflects off a given material. This is important because most solar energy that hits a very reflective surface is not absorbed and therefore does little to warm

Earth. Water in the oceans or on a lake is one of the darkest surfaces, reflecting less than 10% of the incident light, while clouds and snow or ice are among the brightest surfaces, reflecting 70% to 90% of the incident light. When sea ice melts, as it has done in the Arctic Ocean at a disturbing rate over the past decade, the albedo of the area affected changes dramatically, from around 80% down to less than 10%. Much more solar energy is absorbed by the water than by the pre-existing ice, and the temperature increase is amplified. The same applies to ice and snow on land, but the difference in albedo is not as great. When ice and snow on land melt, sea level rises. (Sea level is also rising because the oceans are warming and that increases their volume; see section 13.7). A higher sea level means a larger proportion of the planet is covered with water, and since water has a lower albedo than land, more heat is absorbed and the temperature goes up a little more. Since the last glaciation, sea-level rise has been about 125 m; a huge area that used to be land is now flooded by heat-absorbent seawater. During the current period of anthropogenic climate change, sea level has risen only about 20 cm, and although that doesn't make a big change to albedo, sea-level rise is accelerating.

Most of northern Canada, Alaska, Russia, and Scandinavia has a layer of permafrost that ranges from a few centimeters to hundreds of meters in thickness. Permafrost is a mixture of soil and ice and it also contains a significant amount of trapped organic carbon that is released as CO_2 and CH_4 when the permafrost breaks down. Because the amount of carbon stored in permafrost is in the same order of magnitude as the amount released by burning fossil fuels, this is a feedback mechanism that has the potential to equal or surpass the forcing that has unleashed it. In some polar regions, including northern Canada, permafrost includes methane hydrate, a highly concentrated form of CH_4 trapped in solid form. Breakdown of permafrost releases this CH_4. Even larger reserves of methane hydrate exist on the seafloor, and while it would take significant warming of ocean water down to a depth of hundreds of meters, this too is likely to happen in the future if we don't limit our impact on the climate. There is strong isotopic evidence that the Paleocene–Eocene thermal maximum was caused, at least in part, by a massive release of sea-floor methane hydrate.

There is about 45 times as much carbon in the ocean (as dissolved bicarbonate ions, HCO_3^-) as there is in the atmosphere (as CO_2), and there is a steady exchange of carbon between the two reservoirs (see section 5.5). But the solubility of CO_2 in water decreases as the temperature goes up. In other words, the warmer it gets, the more oceanic bicarbonate that gets transferred to the atmosphere as CO_2. That makes CO_2 solubility another positive feedback mechanism. Vegetation growth responds positively to both increased temperatures and elevated CO_2 levels, and so in general, it represents a negative feedback to climate change because the more the vegetation grows, the more CO_2 is taken from the atmosphere. But it's not quite that simple, because when trees grow bigger and more vigorously, forests become darker (they have lower albedo) so they absorb more heat. Furthermore, climate warming isn't necessarily good for vegetation growth; some areas have become too hot, too dry, or even too wet to support the plant community that was growing there, and it might take centuries for something to replace it successfully. All of these positive (and negative) feedbacks work both ways. For example, during climate cooling, growth of glaciers leads to higher albedos, and formation of permafrost results in storage of carbon that would otherwise have returned quickly to the atmosphere.

Anthropogenic Climate Change

When we talk about anthropogenic climate change, we are generally thinking of the industrial era, which really got going when we started using fossil fuels (coal to begin with, and later oil and natural gas) to drive machinery and trains, and to generate electricity. That was around the middle of the 18th century. The issue with fossil fuels is that they involve burning carbon that was naturally stored in the crust over hundreds of millions of years as part of Earth's process of counteracting the warming Sun.

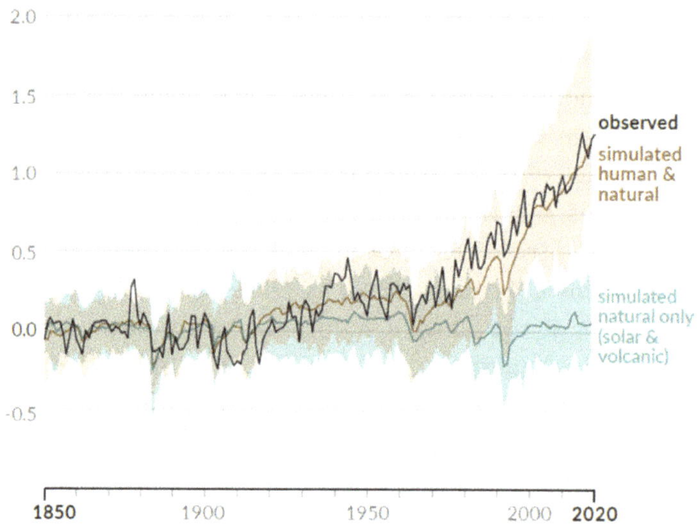

Figure 8.5.2 Changes in global surface temperature over the past 170 years, relative to the average for the period 1850-1900. The black line displays the annual averages of observed global temperatures. The brown line represents the simulated predicted response to both human and natural causes, while the green line represents the simulated predicted responses due to only natural factors. Shaded areas represent the likely range of simulated responses. (From the IPCC 6th Assessment Report Summary for Policymakers; [https://www.ipcc.ch/report/ar6/wg1/downloads/report/IPCC_AR6_WGI_SPM.pdf])

A rapidly rising population, the escalating level of industrialization and mechanization of our lives, and an increasing dependence on fossil fuels have driven the anthropogenic climate change of the past century. The trend of mean global temperatures since 1850 is shown in Figure 8.5.2. For approximately the past 55 years, the temperature has increased at a relatively steady and disturbingly rapid rate, especially compared to past changes. The average temperature now is approximately 1.1°C higher than before industrialization, and two-thirds of this warming has occurred since 1975.

The Intergovernmental Panel on Climate Change (IPCC), established by the United Nations in 1988, is responsible for reviewing the scientific literature on climate change and issuing periodic reports on several topics, including the scientific basis for understanding climate change, our vulnerability to observed and predicted climate changes, and what we can do to limit climate change and minimize its impacts. Figure 8.5.3, from the sixth report of the IPCC, issued in preliminary form in 2021, shows the relative contributions of various greenhouse gases and other factors to current climate forcing, based on the changes from levels that existed in 1750.

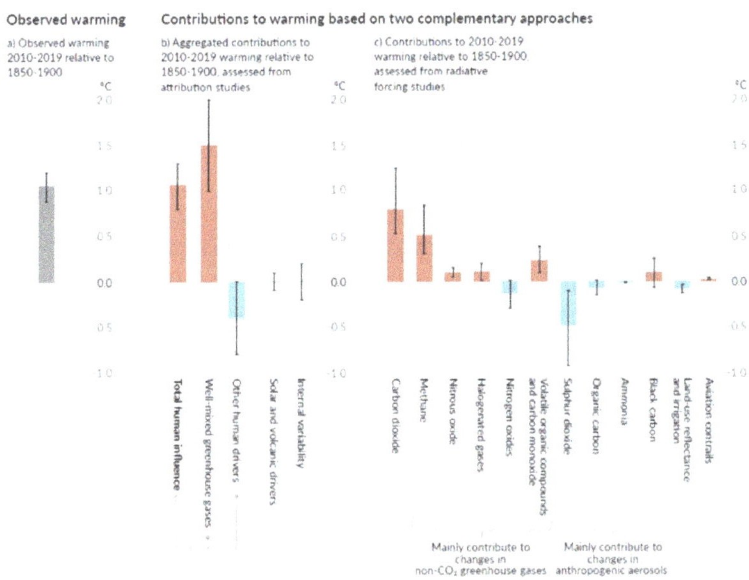

Figure 8.5.3 The relative importance of factors that are contributing to anthropogenic warming (https://www.ipcc.ch/report/ar6/wg1/downloads/report/IPCC_AR6_WGI_SPM.pdf)

The biggest anthropogenic contributor to warming is the emission of CO_2, which accounts for 50% of positive forcing. CH_4 and its atmospheric derivatives (CO_2, H_2O, and O_3) account for 29%, and the halocarbon gases (mostly leaked from air-conditioning appliances) and nitrous oxide (N_2O) (from burning fossils fuels) account for 5% each. Carbon monoxide (CO) (also produced by burning fossil fuels) accounts for 7%, and the volatile organic compounds other than methane (NMVOC) account for 3%. CO_2 emissions come mostly from coal- and gas-fired power stations, motorized vehicles (cars, trucks, and aircraft), and industrial operations (e.g., smelting), and indirectly from forestry. CH_4 emissions come from production of fossil fuels (escape from coal mining and from gas and oil production), livestock farming (mostly beef), landfills, and wetland rice farming. N_2O and CO come mostly from the combustion of fossil fuels. In summary, close to 70% of our current greenhouse gas emissions come from fossil fuel production and use, while most of the rest comes from agriculture and landfills. Figure 8.5.4 shows the IPCC's projections for temperature increases over the next 100 years as a result of these increasing greenhouse gases.

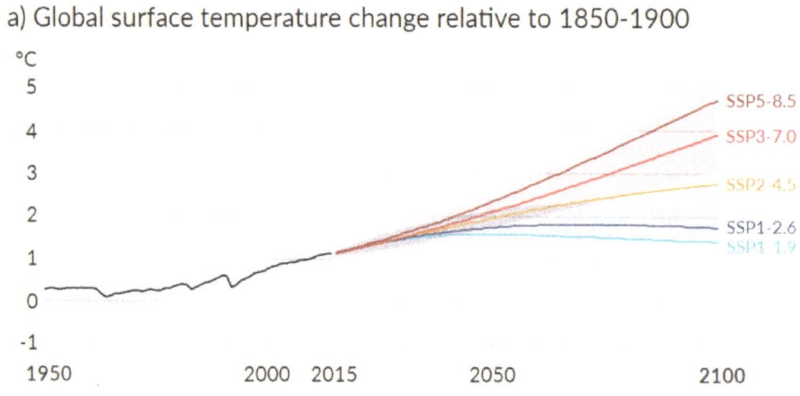

Figure 8.5.4 Projected global temperature increases for the 21st century based on a range of different IPCC scenarios of future political, technological, and emissions variables (https://www.ipcc.ch/report/ar6/wg1/downloads/report/IPCC_AR6_WGI_SPM.pdf).

Impacts of Climate Change

Throughout the past decade, we have all personally witnessed the consequences of climate change. Nevertheless, climatologists face challenges in establishing a direct link between a warming climate and certain weather occurrences, and therefore understandably hesitate to attribute any given event solely to climate change. The most reliable indicators of climate change are those that can be observed over long periods of time, such as the temperature fluctuations depicted in Figure 8.5.2 or the increase in sea levels illustrated in Figure 8.5.5. The sea level has increased by around 20 cm since 1750 due to the combined effects of warmer saltwater, which causes expansion, and the melting of glaciers and other land-based snow and ice.

(melting of sea ice does not contribute directly to sea level rise as it is already floating in the ocean, see section 13.7).

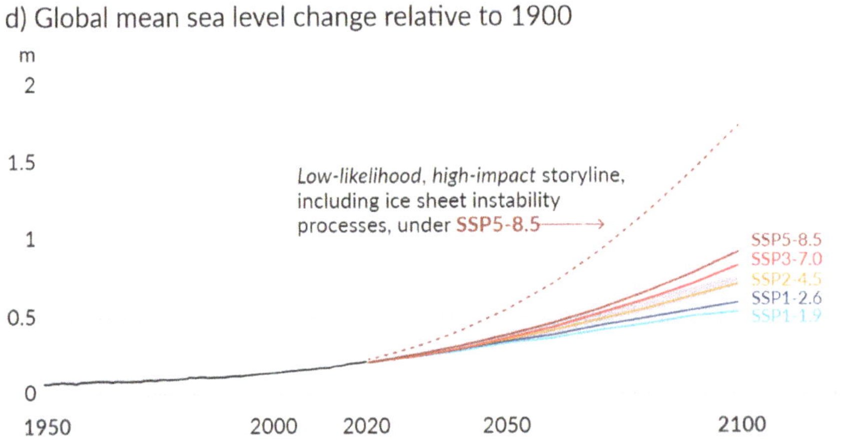

Figure 8.5.5 Current and projected potential sea-level increases to 2100, relative to 1900. Each line represents a different IPCC scenario of future political, technological, and emissions variables (https://www.ipcc.ch/report/ar6/wg1/downloads/report/IPCC_AR6_WGI_SPM.pdf).

Projections for sea level rise to the end of this century vary widely. This is in large part because we do not know which of the above climate change scenarios (Figure 8.5.4) we will most closely follow, but many are in the range from 0.5 m to 2.0 m. One of the problems in predicting sea level rise is that we do not have a strong understanding of how large ice sheets, such as Greenland and Antarctica, will respond to future warming. Another issue is that the oceans don't respond immediately to warming.

For example, with the current amount of warming, we are already committed to a future sea level rise of between 1.3 m and 1.9 m, even if we could stop climate change today. This is because it takes decades to centuries for the existing warming of the atmosphere to be transmitted to depth within the oceans and to exert its full impact on large glaciers. Most of that committed rise would take place over the next century, but some would be delayed longer. And for every decade that the current rates of climate change continue, that number increases by another 0.3 m. In other words, if we don't make changes quickly, by the end of this century we'll be locked into 3 m of future sea level rise.

In a 2008 report, the Organization for Economic Co-operation and Development (OECD) estimated that by 2070 approximately 150 million people living in coastal areas could be at risk of flooding due to the combined effects of sea level rise, increased storm intensity, and land subsidence. The assets at risk (buildings, roads, bridges, ports, etc.) are in the order of $35 trillion ($35,000,000,000,000). Countries with the greatest exposure of population to flooding are China, India, Bangladesh, Vietnam, U.S.A., Japan, and Thailand. Some of the major cities at risk include

Shanghai, Guangzhou, Mumbai, Kolkata, Dhaka, Ho Chi Minh City, Tokyo, Miami, and New York.

One of the other risks for coastal populations, besides sea level rise, is that climate warming is also associated with an increase in the intensity of tropical storms (e.g., hurricanes or typhoons; see section 8.4), which almost always bring serious flooding from intense rain and storm surges. Some recent examples are New Orleans in 2005 with Hurricane Katrina, and New Jersey and New York in 2012 with Hurricane Sandy. Tropical storms get their energy from the evaporation of warm seawater in tropical regions. In the Atlantic Ocean, this takes place between 8° and 20° N in the summer. Figure 8.5.6 shows the variations in the sea-surface temperature (SST) of the tropical Atlantic Ocean (in blue) versus the amount of power represented by Atlantic hurricanes between 1950 and 2008 (in red). Not only has the overall intensity of Atlantic hurricanes increased with the warming since 1975, but the correlation between hurricanes and sea-surface temperatures is very strong over that time period.

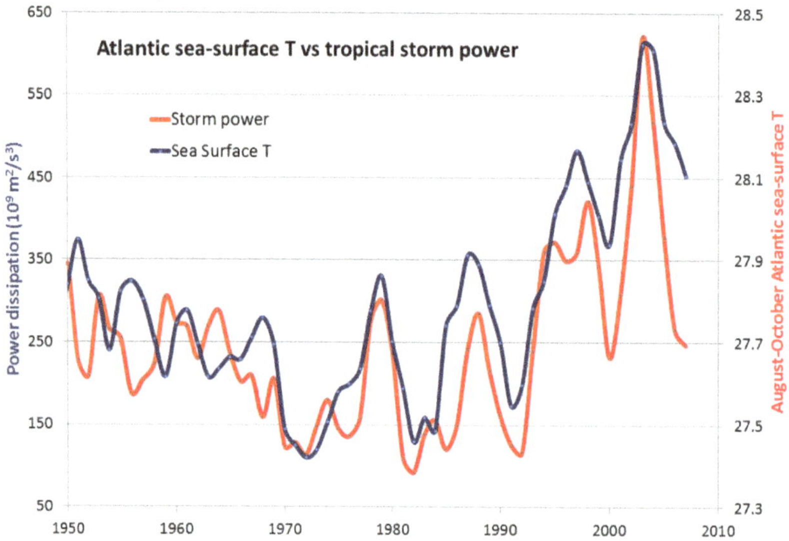

Figure 8.5.6 Relationship between Atlantic tropical storm cumulative annual intensity and Atlantic sea-surface temperatures (Steven Earle, "Physical Geology", by SE from data at: http://wind.mit.edu/~emanuel/Papers_data_graphics.htm).

Climate change has been observed to cause the expansion of disease and pest ranges, particularly those produced or transmitted by insects, into temperate zones. West Nile virus and Lyme disease are two existing instances of diseases that directly impact the population of North America. Additionally, dengue fever has the potential to become a concern in the future, as it was officially recognized as a "nationally notifiable condition" in the United States in 2010. In the months of July and August 2010, a substantial heat wave impacted western Russia, particularly the region southeast of Moscow. Scientists have attributed climate change as a significant influence in this occurrence. The temperatures rose to more than 40°C, above the typical levels by up to 12°C throughout a large region, while wildfires spread in various areas of the country. The heat and respiratory difficulties caused by the fires are responsible for more than 55,000 deaths. Figure 8.5.7 provides a concise overview of the effects of climate change on natural disasters. Floods and storms are the primary categories of climate-related disasters, however the health consequences of severe temperatures are also more worrisome. Between 1971 and 1980, severe temperatures ranked as the fifth most frequent natural disasters. However, between 2001 and 2010, they rose to become the third most prevalent.

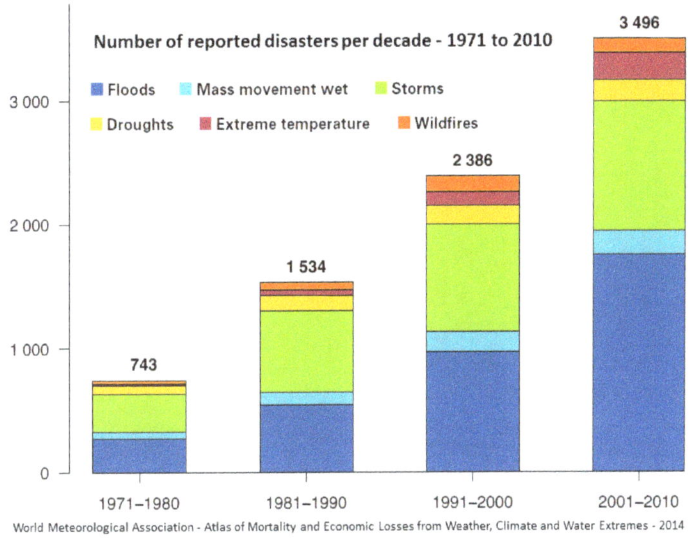

Figure 8.5.7 Numbers of various types of disasters between 1971 and 2010 (From WMO atlas of mortality and economic Losses from weather, climate and water extremes, 2014).

9
Chapter
WORLD OCEAN CIRCULATION

INTRODUCTION: AN OVERVIEW

The movement of ocean waters is perpetual, encompassing large-scale surface currents, vertical turnover driven by differences in density, and the presence of small rotating eddies. Oceanographers possess a wide range of advanced instruments to measure ocean currents. However, occasionally, unexpected incidents can also contribute to our comprehension of ocean circulation. An illustrative instance is the incident involving the container ship Ever Laurel, which was in route from Hong Kong to Tacoma, Washington, in January 1992. During a storm in the midst of the Pacific, a total of 12 containers were swept overboard. One of the containers held more than 28,000 plastic bath toys, which were discharged into the ocean upon contact with the water. After a period of ten months, the bath toys started appearing on the shore, initially near Sitka, Alaska, and then at other locations along the Alaskan coast.

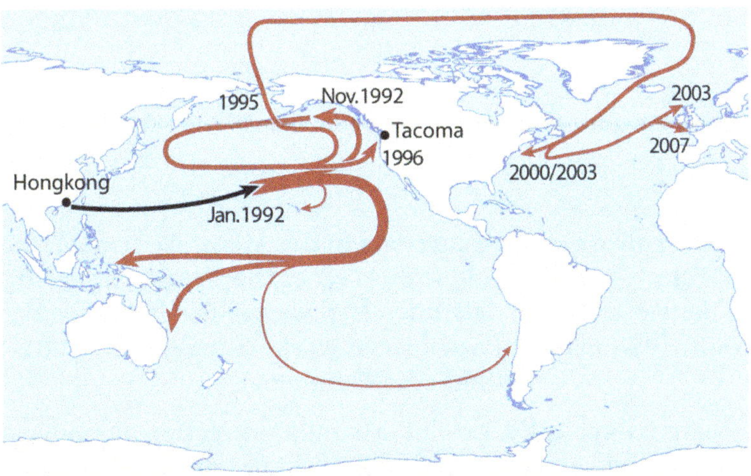

Figure 9.1 Paths taken by the floating toys lost from the Ever Laurel in 1992, showing the years in which toys were found in different locations (NordNordWest [CC BY-SA 3.0] via Wikimedia Commons).

By 1996, they were also found in Washington. During the following twenty years, several toys were transported to the Pacific shores of South America and Australia. Additionally, some toys were discovered in the Arctic ice, and a few even managed to pass through the Arctic and reach the North Atlantic, where they washed up in Newfoundland and Scotland (Figure 9.1). Several thousand toys remain adrift in the central North Pacific, and the trajectories of these toys have facilitated the study of major ocean surface currents by oceanographers.

9.1 SURFACE GYRES

The previous chapter provided an analysis of the primary wind patterns seen on Earth. The principal ocean surface currents are created by the prevailing winds that blow across the water surface. However, an only 2% of the wind energy is effectively transmitted to the water, resulting in a current of merely 1 knot when there is a 50 knot wind. Moreover, surface currents caused by wind exclusively impact the uppermost 100-200m of water, indicating that surface currents encompass merely around 10% of the total volume of ocean water worldwide. Section 9.8 will focus on the analysis of deep, thermohaline circulation, which affects approximately 90% of the water in the ocean.

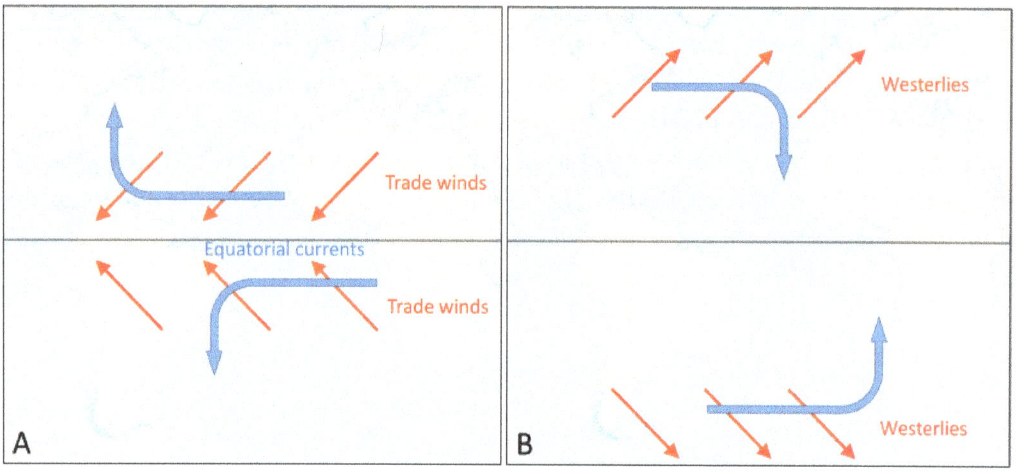

Figure 9.1.1 Generalized surface currents in the Atlantic Ocean. A) Surface water moving at 45° relative to the trade winds create the westward flowing equatorial currents. B) Between 30-60° latitude, the westerlies form eastward flowing surface currents (PW, map by Catrin (Own work, Using GMT) [CC BY-SA 3.0], via Wikimedia Commons).

Surface currents typically exhibit a similar direction of movement as the winds that generated them. Due to the Coriolis effect, the surface currents deviate by approximately 45 degrees from the direction of the wind. In the Northern Hemisphere, the deviation is towards the right, whereas in the Southern Hemisphere, it is towards the left. This results in a prevailing circulation pattern in which surface currents move from east to west between the equator and 30o latitude, from west to east between 30o and 60o, and from east to west between 60o and the poles (Figure 9.1.1).

The trade winds generate the equatorial currents, specifically the North Equatorial and South Equatorial currents, which move from east to west along the equator. In the absence of continents, these surface currents would circulate around the Earth in a continuous path that runs parallel to the equator. Nevertheless, the existence of the continents hinders this unrestricted movement. Equatorial currents, upon reaching the continents, are redirected and deviated from the equator due to the Coriolis Effect. In the Northern Hemisphere, they are deflected to the right, whereas in the Southern Hemisphere, they are deflected to the left. Subsequently, these currents transform into western boundary currents, which flow adjacent to the western edge of the ocean basin (namely, the eastern shores of the continents). As these currents originate from the equator, they are warm water currents that transport warm water to the higher latitudes and disperse heat across the ocean.

Simultaneously, within the latitudes of 30-60o, the westerlies transport surface water in an eastward direction. The Coriolis Effect and the continents cause the currents to deviate towards the equator, resulting in the formation of eastern boundary currents along the eastern edges of the ocean basins. The currents originate from regions located at high latitudes, resulting in the transportation of cold water towards lower latitudes. Collectively, these currents merge to generate extensive circular patterns of surface circulation known as gyres. In the Northern Hemisphere, the gyres exhibit a clockwise rotation, but in the Southern Hemisphere, the gyres display a counterclockwise rotation.

There are five primary gyres in the world's oceans: the North Atlantic, South Atlantic, North Pacific, South Pacific, and Indian Ocean gyres (Figure 9.1.2). The North Pacific gyre consists of the North Equatorial Current along its southern edge, which transforms into the Kuroshio Current (also known as the Japan Current), carrying warm water northward into Japan. The Kuroshio merges with the North Pacific Current and proceeds eastward towards North America, transforming into the California Current to form a circular oceanic current system. The North Atlantic gyre is created by the convergence of the North Equatorial Current with the Gulf Stream around the eastern coast of the United States. The Gulf Stream converges with the North Atlantic Current, facilitating the transportation of water towards Europe. Subsequently, this water transforms into the Canary Current as it flows southward and unites with the North Equatorial Current.

In the vicinity of Antarctica, the circulation exhibits certain variations. The absence of significant land masses between 50-60o south allows the surface current generated by the westerly winds to circulate around the entire Earth, forming the Antarctic Circumpolar Current (ACC) or West Wind Drift (WWD). This current runs from west to east (Figure 9.1.2). The Antarctic Circumpolar Current is the sole current that links all the primary ocean basins, and it has the distinction of being the most extensive surface current on Earth in terms of water volume it carries. At latitudes above 60 degrees, the dominant winds are known as the polar easterlies. These winds generate a current that flows from east to west along the perimeter of the Antarctic continent, referred to as the East Wind Drift or the Antarctic Coastal Current.

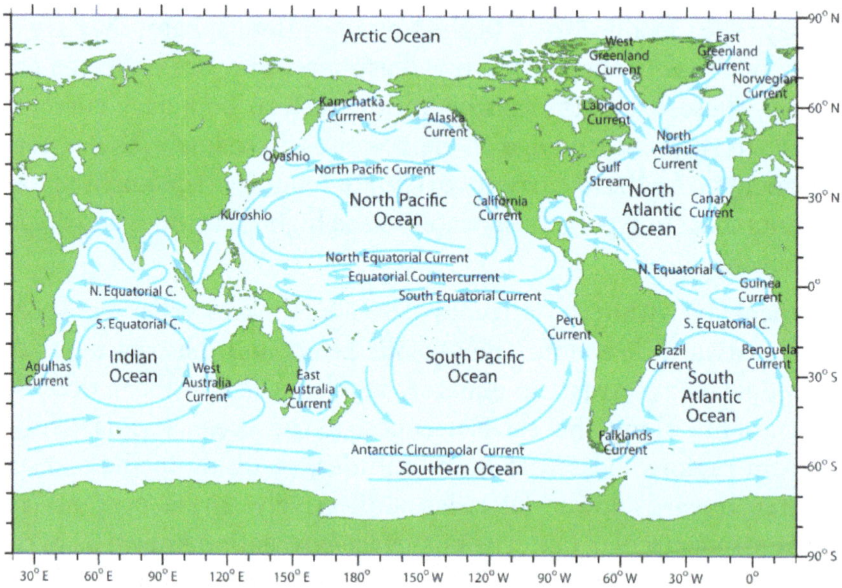

Figure 9.1.2 The major global surface currents (http://www.seos-project.eu/modules/oceancurrents/oceancurrents-c02-p01.html, Source: NOC; CC BY-NC-SA 2.0).

The Antarctic Circumpolar Current serves as the southern demarcation for all of the gyres in the Southern Hemisphere. The Antarctic Circumpolar Current (ACC) transforms into the Peru Current, often referred to as the Humboldt Current, when it travels through the South Pacific gyre. The Peru Current then proceeds northward along the western coast of South America until it merges with the South Equatorial Current. The South Equatorial Current travels in a southerly direction and is known as the East Australia Current. It then completes the round pattern of ocean currents, known as a gyre, by joining with the Antarctic Circumpolar Current (ACC). The South Atlantic gyre consists of four major currents: the South Equatorial Current, the Brazil Current, the Antarctic Circumpolar Current (ACC), and the Benguela Current. The Indian gyre is comprised of four main currents: the Antarctic Circumpolar Current (ACC), the West Australia Current, the South Equatorial Current, and the Agulhas Current.

Only a portion of the equatorial water, which is carried westward by the trade winds and reaches the continents, is actually transferred to higher latitudes in the gyres. This is due to the fact that the Coriolis Effect is least pronounced along the equator. Alternatively, a portion of the water accumulates near the western boundary of the ocean and then moves towards the east as a result of gravitational forces. This phenomenon gives rise to slender Equatorial Countercurrents that exist between the North and South Equatorial Currents (as depicted in Figure 9.1.2). A portion of this water also migrates towards the east as equatorial undercurrents, which travel at depths ranging from 50 to 200 meters, beneath the Equatorial Currents. The Lomonosov Current is the term used to refer to these undercurrents in the Atlantic, while the Cromwell Current is the one used in the Pacific.

9.2 THE GULF STREAM

The dominant surface current along the eastern seaboard of the United States is known as the Gulf Stream. This current was initially charted by Benjamin Franklin during the 18th century (see Figure 9.2.1). As a powerful and swift oceanic current, it significantly decreased the duration of voyages for ships heading from the United States to Europe. Sailors would employ thermometers to identify its warm water and remain within the current.

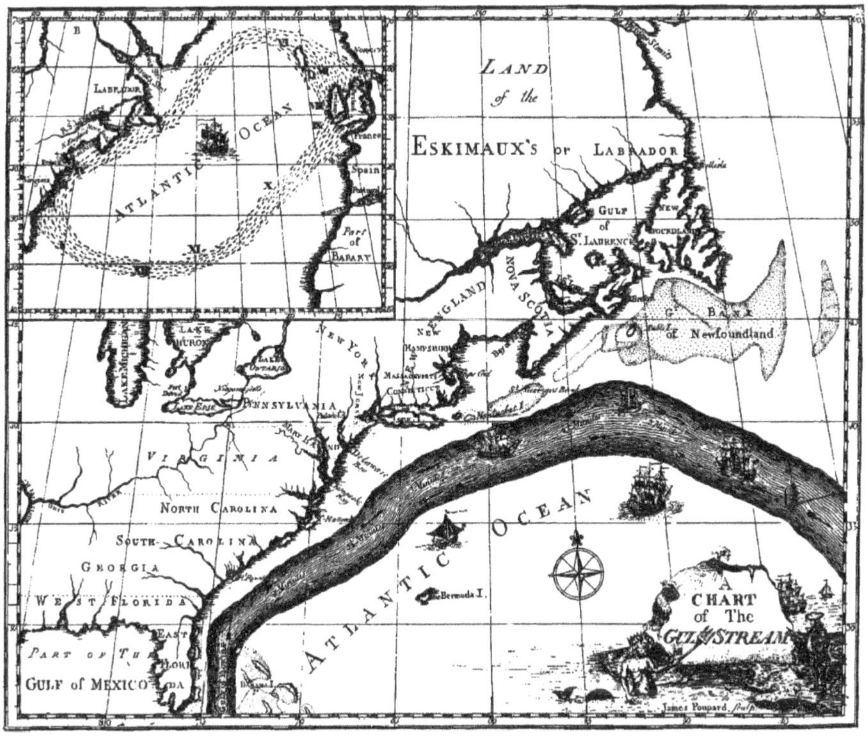

Figure 9.2.1 Benjamin Franklin's original map of the Gulf Stream (Public domain, via Wikimedia Commons).

The Gulf Stream is formed from the convergence of the North Atlantic Equatorial Current bringing tropical water from the east, and the Florida Current that brings warm water from the Gulf of Mexico. The Gulf Stream takes this warm water and transports it northwards along the U.S. east coast (Figure 9.2.2). As a western boundary current, the Gulf Stream experiences western intensification (section 9.4), making the current narrow (50-100 km wide), deep (to depths of 1.5 km) and fast. With an average speed of 6.4 km/hr, and a maximum speed of about 9 km/hr, it is the fastest current in the world ocean. It also transports huge amounts of water, more than 100 times greater than the combined flow of all of the rivers on Earth.

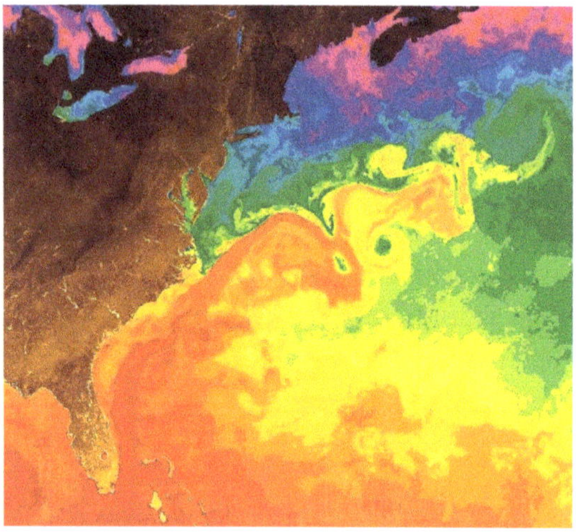

Figure 9.2.2 Sea surface temperature map illustrating the Gulf Stream. Warmer water is shown in red, colder water in blue and violet. Meanders and eddies are visible where the current moves towards the northeast (NASA, Public Domain via Wikimedia Commons).

As the Gulf Stream approaches Canada, the current becomes wider and slower as the flow dissipates and it encounters the cold Labrador Current moving in from the north. At this point, the current begins to meander, or change from a fast, straight flow to a slower, looping current (Figure 9.2.2). Often these meanders loop so much that they pinch off and form large rotating water masses called **rings** or **eddies**, that separate from the Gulf Stream. If an eddy pinches off from the north side of the Gulf Stream, it entraps a mass of warm water and moves it north into the surrounding cold water of the North Atlantic. These **warm core rings** are shallow, bowl-shaped water masses about 1 km deep, and about 100 km across, that rotate clockwise as they carry warm water in to the North Atlantic (Figure 9.2.3). If the meanders pinch off at the southern boundary of the Gulf Stream, they form **cold core rings** that rotate counterclockwise and move to the south. Cold core rings are cone-shaped water masses extending down to over 3.5 km deep, and may be over 500 km wide at the surface.

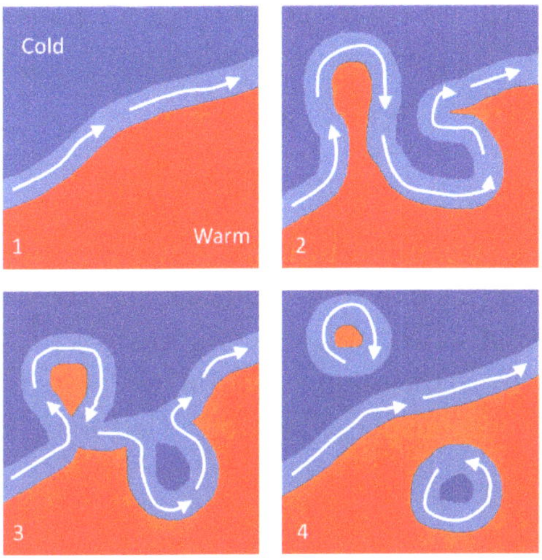

Figure 9.2.3 Formation of warm and cold core rings from meanders in the Gulf Stream. As the Gulf Stream flows to the northeast (1), it starts to meander as it slows, forming warm or cold water extensions on either side of the current (2). If the meanders pinch off the extensions, they trap pockets of warm or cold water (3), that can separate from the Gulf Stream and travel north or south. Warm core rings rotate clockwise, while cold core rings rotate counterclockwise (4) (PW).

After the Gulf Stream meets the cold Labrador Current, it joins the North Atlantic Current, which transports the warm water towards Europe, where it moderates the European climate. It is estimated that Northern Europe is up to 9° C warmer than expected because of the Gulf Stream, and the warm water helps to keep many northern European ports ice-free in the winter.

Figure 9.2.4 Map of the Sargasso Sea (left), and closeup photo of Sargassum algae (right) (Map by USFWS; photo by Bogdan Giuşcă; Public Domain via Wikimedia Commons).

In the east, the Gulf Stream merges into the Sargasso Sea, which is the area of the ocean within the rotation center of the North Atlantic gyre. The Sargasso Sea gets its name from the large floating mats of the marine algae *Sargassum* that are abundant on the surface (Figure 9.2.4). These *Sargassum* mats may play an important role in the early life stages of sea turtles, who may live and feed within the algae for many years before reaching adulthood.

9.3 THE EKMAN SPIRAL AND GEOSTROPHIC FLOW

Ekman spiral

Winds blowing over the ocean are ultimately what create the surface currents. However, not all of the water moved by the surface currents is transported in the same direction. The Coriolis Effect causes the surface water to move in a direction about 45 degrees offset from the wind direction, with the deflection to the right of the wind in the Northern Hemisphere and to the left in the Southern Hemisphere.

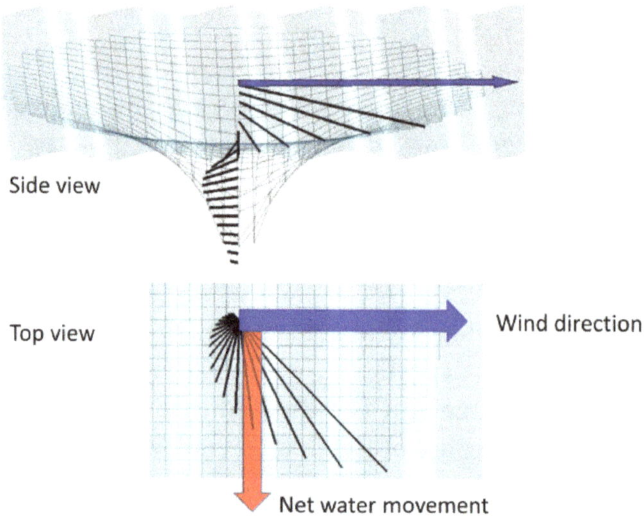

Figure 9.3.1 The Ekman spiral, shown for the Northern Hemisphere. Wind blowing over the water (blue arrow) creates a surface current 45° offset from the wind. Each successive layer of water is moved and deflected by the layer above, creating a spiraling pattern of water movement that diminishes with depth. The net movement of the water within the spiral is 90° relative to the wind direction (red arrow) (Modified by PW from Ekman spiral By Schlusenbach (unbekannt) [Public domain], via Wikimedia Commons).

The frictional movement of the topmost layer of water sets in motion the layer directly underneath it, which then sets in motion the next layer under that, and so on as the water gets deeper. Some energy is lost in each transition, so each successive layer of water will not move as far as the layer above it; in other words, there is decreasing energy with increasing depth. But at the same time, the Coriolis Effect deflects each layer relative to the layer above it (again, to the right in the Northern Hemisphere and to the left in the Southern Hemisphere). The movement of the successive layers therefore creates a spiraling pattern of water motion called the **Ekman spiral**, which usually penetrates to about 100 m deep before the motion ceases. If the magnitudes and directions of the movements of each layer are added together, the result is that the net movement of the upper 100 m of the water column is 90° relative to the original wind direction (90° to the right of the wind in the Northern Hemisphere, and 90° to the left in the Southern Hemisphere). This net water movement is called **Ekman transport** (Figure 9.3.1).

Geostrophic flow

Gyre rotation is dependent on the wind and the Coriolis Effect impacting the surface currents (see section 9.1). But the rotation is also affected by movement below the surface due to Ekman transport. In the Northern Hemisphere, as the gyre rotates clockwise the net movement of the Ekman transport is 90° to the right of the wind; in other words, towards the center of the gyre. The Ekman transport piles up water in the center of the gyre, making the water level higher in the gyre center than on the edges of the gyres. This pile of water then has a tendency to flow back "downhill" due to gravity. As the water flows "downhill" away from the gyre center, it is deflected to the right by the Coriolis force. This results in a clockwise current around the central "hill" called **geostrophic flow**, which moves in the same direction as the rotating gyre. Water is thus pushed into the "hill" by Ekman transport, and away from the "hill" by gravity, with both flows modified by the Coriolis Effect to create the rotation. As with the gyres, geostrophic flow is clockwise in the Northern Hemisphere, and counterclockwise in the Southern Hemisphere.

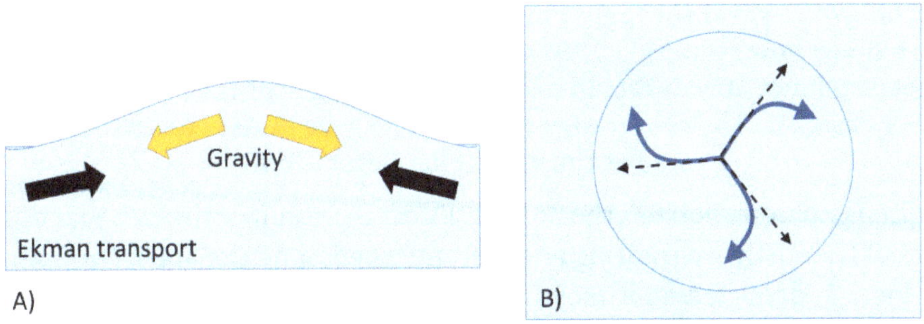

Figure 9.3.2 Geostrophic flow in the Northern Hemisphere. A) Ekman transport moves water into the middle of the gyre, where it "piles up." Gravity causes the water to flow back "downhill." B) Viewed from above, as the water in the center flows "downhill" (dotted arrows) the Coriolis force deflects the movement to the right (solid arrows), causing the system to rotate clockwise (PW).

Most major surface currents are a combination of wind-driven and geostrophic currents. Since winds can be variable, geostrophic flow ensure that the gyre currents keep moving at a fairly constant rate even when the wind dies down. The larger the area, and the higher the slope, the longer the geostrophic flow will continue to move and power the gyre after the wind subsides.

9.4 WESTERN INTENSIFICATION

In both hemispheres, the currents making up the western side of the gyre are much more intense than the currents on the eastern side. In other words, the currents off of the east coast of the continents are more intense than currents off of the west coast of the continents. This phenomenon is known as western intensification, and once again it is due to the Coriolis Effect.

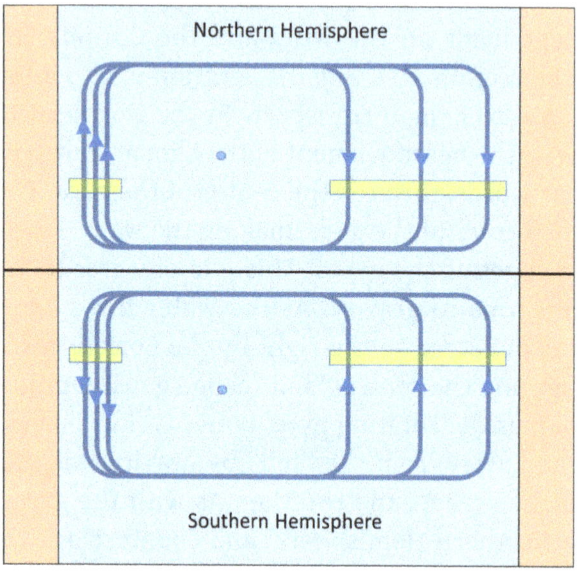

Figure 9.4.1 Western intensification. In both hemispheres, currents on the western side of the gyres travel through a much narrower area than the currents on the eastern side (yellow rectangles). To move the same volume of water through each side, western boundary currents are faster, deeper, and narrower than eastern boundary currents. The center of rotation of the gyre is also closer to the western side of the gyre (blue dots) (PW).

As discussed in section 8.2, the Coriolis Effect is a result of the fact that different latitudes of the Earth are rotating at different speeds, and the apparent path taken by an object is deflected as it moves between areas of different rotation speeds. The greater the change in rotation speed, the stronger the Coriolis force. At the poles, the speed of rotation is 0 km/hr. The speed increases to about 800 km/hr at 60° latitude, 1400 km/hr at 30° latitude, and 1600 km/hr at the equator. Therefore there is an 800 km/hr difference between 60° and 90° latitude, while there is only a 200 km/hr difference between the equator and 30°. Thus the speed of Earth's rotation changes more quickly with latitude near the poles than at the equator, making the Coriolis force strongest near the poles and weakest at the equator.

The high latitude surface currents of the major gyres experience a strong Coriolis force due to their proximity to the poles. As the currents move eastward, the strong Coriolis force begins to deflect the currents towards the equator relatively early. The currents on the eastern side of the gyre are therefore spread out over a wide area as they move towards the equator (Figure 9.4.1). Near the equator, the westward flowing currents experience a much weaker Coriolis force, so their deflection does not happen until the current is all the way over to the western side of the ocean basin. These western currents must therefore move through a much narrower area (Figure 9.4.1). This imbalance means that the center of rotation of the gyre is not in the center of the ocean basins, but it closer to the western side of the gyre.

The same volume of water must pass through both the east and west sides of the gyre. In the western gyre currents, that volume is passing through a narrower area, so the current must travel faster in order to transport the same amount of water in the same amount of time. On the eastern side of the gyre the current is much wider, so the flow is slower. A simple analogy is the water flowing from a garden hose. You can make the water flow from the hose much faster and more strongly by covering part of the opening with your thumb. The same amount of water is exiting the hose whether the opening is covered or uncovered, but to get that water through the covered opening the flow has to be much faster and stronger. In the same way, western boundary currents are not only faster, but also deeper than eastern boundary currents, as they move the same volume through a narrower space. For example, the Kuroshio Current in the western Pacific is around 15 times faster, 20 times narrower, and 5 times deeper than the California Current in the eastern Pacific.

9.5 CURRENTS, UPWELLING AND DOWN-WELLING

The movement of surface currents also plays a role in the vertical movements of deeper water, mixing the upper water column. **Upwelling** is the process that brings deeper water to the surface, and its major significance is that it brings nutrient-rich deep water to the nutrient-deprived surface, stimulating primary production (see section 7.3). **Downwelling** is where surface water is forced downwards, where it may deliver oxygen to deeper water. Downwelling leads to reduced productivity, as it extends the depth of the nutrient-limited layer.

Upwelling occurs where surface currents are diverging, or moving away from each other. As the surface waters diverge, deeper water must be brought to the surface to replace it, creating upwelling zones. The upwelled water is cold and rich in nutrients, leading to high productivity. Many of the most productive regions on Earth are found in upwelling zones. In the equatorial Pacific, the trade winds blow the North and South Equatorial Currents towards the west, while Ekman transport causes the upper layers to move to the north and south in their respective hemispheres. This creates a divergence zone, and a region of upwelling and high productivity (Figure 9.5.1).

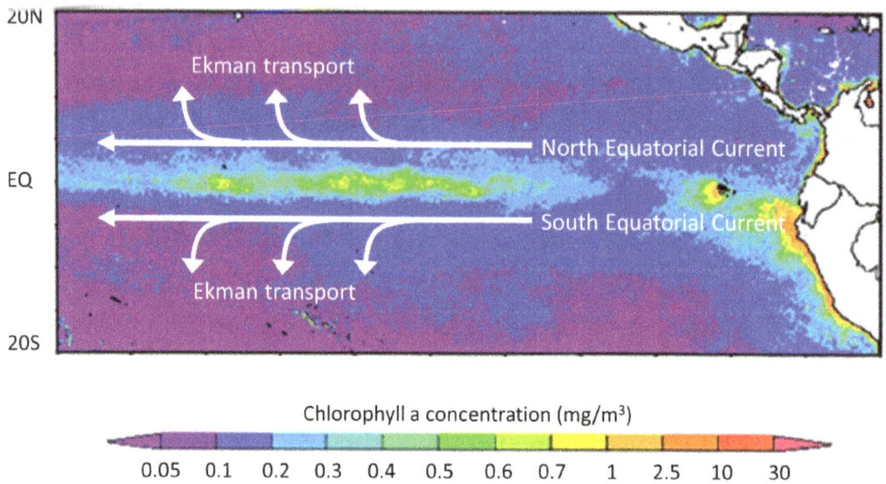

Figure 9.5.1 Equatorial upwelling and increased productivity as a result of divergence between the north and south equatorial currents (Modified by PW from image by NASA [Public domain], via Wikimedia Commons).

A similar process occurs near the Antarctic continent, creating one of the most productive regions on Earth, the Antarctic divergence. In this case, the West Wind Drift (Antarctic Circumpolar Current) is flowing parallel to, but in the opposite direction of the East Wind Drift. With both currents occurring in the Southern Hemisphere, Ekman transport will be to the left, so the eastward-flowing West Wind Drift water will be transported to the north, and the westward-flowing East Wind Drift water will be transported to the south, creating a highly productive divergence zone (Figure 9.5.2).

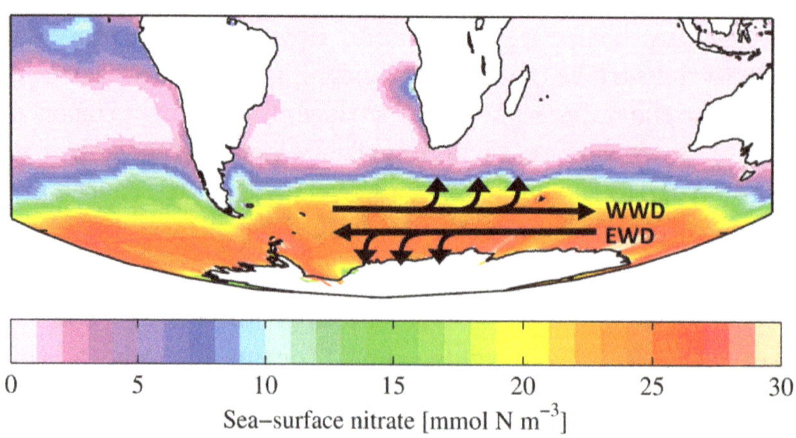

Figure 9.5.2 High nutrient levels in the Antarctic divergence zone, as a result of the diverging West Wind Drift and East Wind Drift currents creating strong upwelling (Modified by PW from Plumbago (Own work) [CC BY-SA 3.0], via Wikimedia Commons).

Downwelling occurs where surface currents converge. The converging water has nowhere to go but down, so the surface water sinks. Since surface water is usually low in nutrients, downwelling leads to low productivity zones. An example of a downwelling region is off of the Labrador coast in Canada, where the Gulf Stream, Labrador, and East Greenland Currents converge.

Coastal Upwelling

Upwelling and downwelling also occur along coasts, when winds move water towards or away from the coastline. Surface water moving away from land leads to upwelling, while downwelling occurs when surface water moves towards the land. Historically, some of the most productive commercial fishing grounds have been associated with coastal upwelling. Along the coast of California, the local prevailing winds blow towards the south. Ekman transport moves the surface layer 90° to the right of the wind, meaning the net Ekman transport is in an offshore direction. The water displaced near the coast is replaced by cold, nutrient-rich deeper water that is brought to the surface through upwelling, leading to high productivity (Figure 9.5.3).

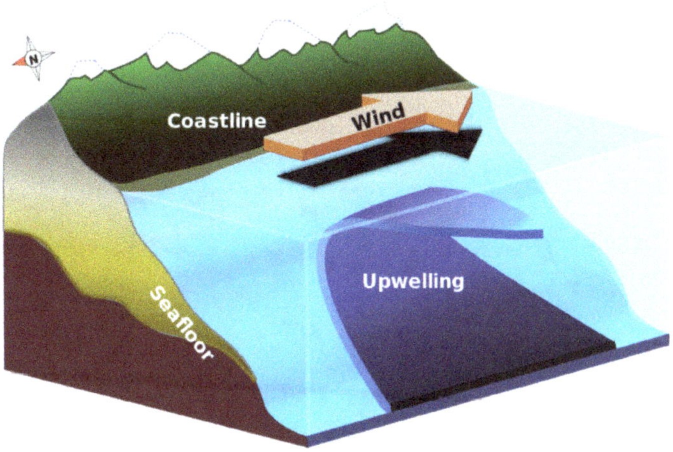

Figure 9.5.3 Coastal upwelling. As wind blows along a coastline, Ekman transport moves the surface layer in a direction 90° to the wind. In this figure, it is to the right of the wind, indicating a Northern Hemisphere location. As surface water moves offshore it it replaced by upwelled deeper water (By Lichtspiel [Public domain], via Wikimedia Commons).

The same process happens off of the coast of Peru, which for a long time had the world's largest commercial fishery. Winds along the Peruvian coast blow towards the north, and since Peru is in the Southern Hemisphere, the Ekman transport is 90° to the left of the wind, which causes the surface water to move offshore and leads to upwelling and productivity. In any coastal upwelling location, if the winds reverse, surface water moves towards the shore and downwelling is the result.

Upwelling can also occur due to geological features of the ocean floor. For example, as deep water currents encounter seamounts or other raised features, the water is forced upwards, bringing nutrient-rich water to the surface. This helps explain why productivity is often high in the water over seamounts.

9.6 EL NIÑO AND LA NIÑA

As we saw in the previous section, coastal upwelling off of Peru makes that region one of the world's most productive fishing grounds. But every so often, the conditions in the region are very different. Every few years, the cold, nutrient-rich water is replaced by unusually warm water that is low in nutrients, leading to a decline in fish populations. In addition, the normally dry areas receive lots of rain. Because this phenomenon occurs in the northern winter close to Christmas, it is called **El Niño** (the child). More formally, the event is referred to as **El Niño-Southern Oscillation (ENSO)**. The Southern Oscillation portion refers to the fluctuating atmospheric conditions that lead to the localized ocean warming of El Niño. While the exact reasons for the oscillation events are unclear, it is easier to understand how they lead to an El Niño.

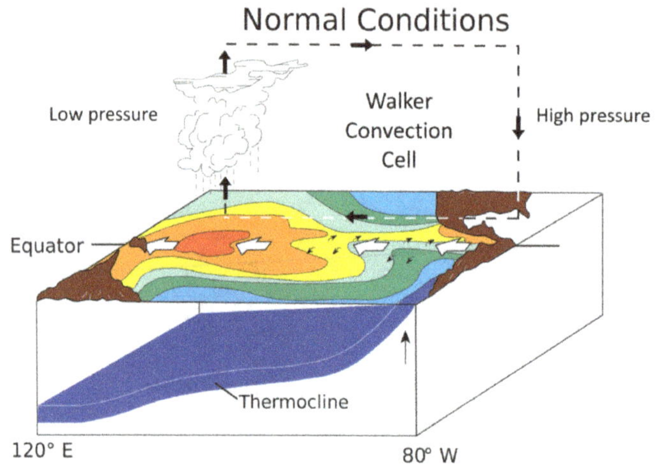

Figure 9.6.1 Normal conditions in the equatorial Pacific. Low pressure in the western Pacific and high pressure in the eastern Pacific cause the trade winds to move surface water to the west, leading to upwelling and a shallow thermocline near South America (Modified by PW from Fred the Oyster (Own work) [Public domain], via Wikimedia Commons).

Under normal conditions in the equatorial Pacific, the trade winds blow towards the west, moving large amounts of warm surface water towards the western Pacific around Southeast Asia. As the surface water moves west, it is replaced by cold, nutrient-rich deep water through upwelling (Figure 9.6.1). The coastal upwelling leads to a shallow thermocline in the eastern Pacific. In terms of atmospheric conditions, the trade winds are part of a convection cell called the Walker Cell.

There is low pressure over the western Pacific, leading to rising moist air and significant precipitation in the region. In the eastern Pacific near South America, there is high pressure, leading to drier conditions (Figure 9.6.1).

During an El Niño-Southern Oscillation, the high pressure system over the eastern Pacific diminishes, so the trade winds are weakened, or in extreme cases will even reverse. When this happens, warm surface water begins to flow east across the Pacific towards South America (Figure 9.6.2), warming the coastal South American water by up to 8° C in strong ENSO years. This influx of low density warm water deepens the thermocline and prevents upwelling, which dramatically reduces productivity and can devastate populations of fish and other marine life.

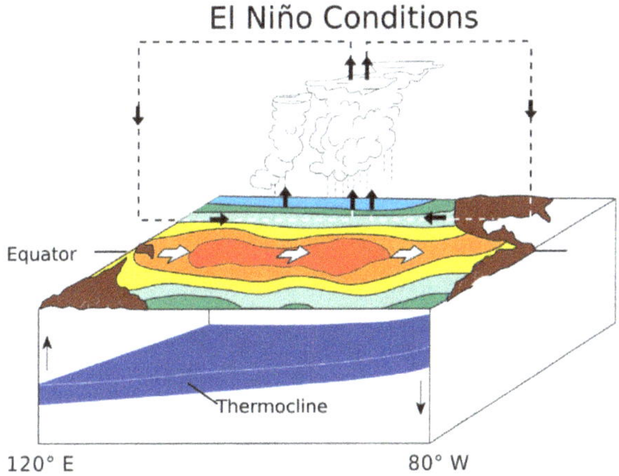

Figure 9.6.2 El Niño conditions in the equatorial Pacific. Weakening or reversal of the trade winds transport warm surface water eastward towards South America, disrupting coastal upwelling (Public domain, via Wikimedia Commons).

The low pressure system in the western Pacific is substituted by a high pressure system in the sky, resulting in the emergence of dry or even drought conditions in Southeast Asia and Australia. The low pressure system traverses eastward across the Pacific, with the possibility of extending its reach to South America during years characterized by strong El Niño events. The presence of a low pressure system in the eastern Pacific results in abundant rainfall and subsequent flooding in South America (Figure 9.6.2). However, the impacts of El Niño are not solely confined to the Pacific region; it has the ability to affect weather patterns worldwide (see to the information provided below).

Due to the periodic nature of the Southern Oscillation, the eastern Pacific is not exclusively exposed to abnormally high temperatures. La Niña episodes are characterized by periods of anomalously cold water in the region. During a La Niña event, the trade winds exhibit exceptional strength, resulting in heightened upwelling and the movement of deep, cold water towards the surface (as shown in Figure 9.6.3). La Niña has contrasting impacts to El Niño, resulting in cooler and more humid weather in the northwestern regions of the United States and Canada, while the southeastern US experiences less rainfall than usual. During El Niño occurrences, the monsoon seasons in Asia see reduced rainfall, while during La Niña events, they receive increased precipitation.

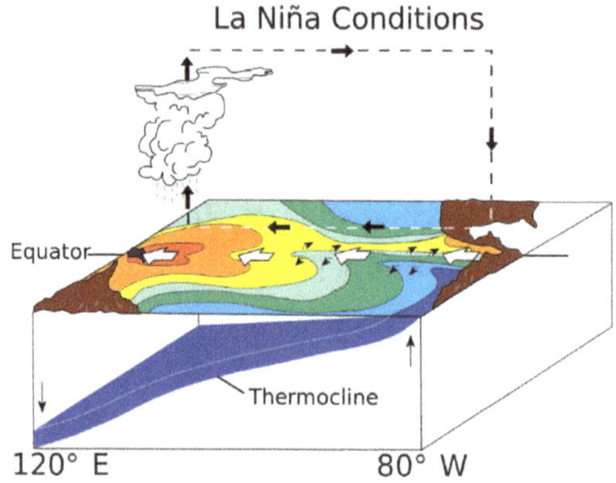

Figure 9.6.3 La Niña conditions. Stronger trade winds promote more intense upwelling in the eastern Pacific, leading to cooler than usual water temperatures (Public domain, via Wikimedia Commons).

El Niño and La Niña occurrences have an alternating pattern, however the occurrence of one does not necessarily imply that the other will inevitably ensue. El Niños typically have a frequency of approximately 2-7 years, and the duration of each event can range from a few months to a year or longer. While the specific causes and timing of ENSO episodes remain unclear, we can predict their occurrence by observing many oceanic and atmospheric factors that constitute the Multivariate ENSO Index (MEI). Analysis of the Multivariate ENSO Index (MEI) over a period of time reveals the recurring pattern of El Niño-Southern Oscillation (ENSO) events. (Figure 9.6.4).

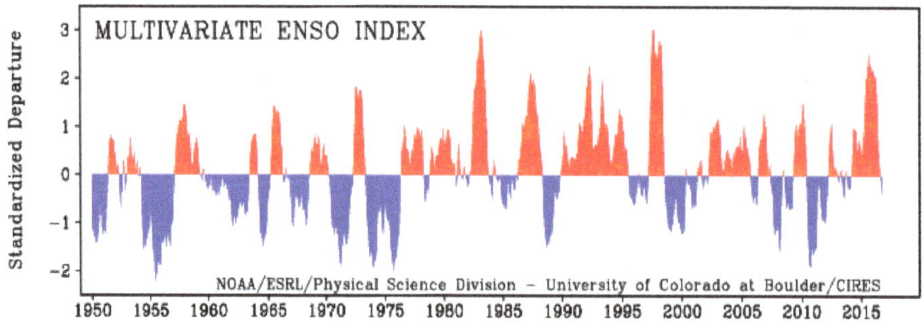

Figure 9.6.4 Multivariate ENSO Index over time. Positive (red) values indicate warmer than normal conditions, while negative (blue) values represent conditions that are cooler than average. The greater the deviation from zero, the stronger the event. Note the intense El Niños in 1983, 1997-1998, and 2015 (NOAA, https://www.esrl.noaa.gov/psd/enso/mei/).

Figure 9.6.5 shows a comparison of sea surface temperatures in the equatorial Pacific during normal, El Niño, and La Niña periods.

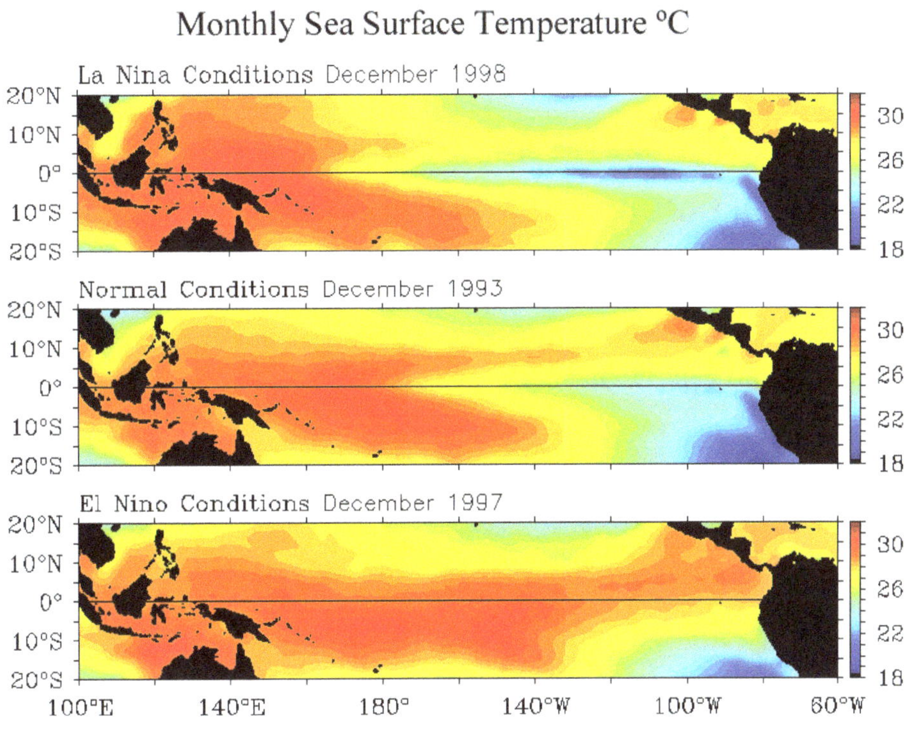

TAO Project Office/PMEL/NOAA

Figure 9.6.5 Comparison of mean December sea surface temperatures in the equatorial Pacific during La Niña (top), normal (middle), and El Niño (bottom) conditions (NOAA, http://www.pmel.noaa.gov/elnino/sites/default/files/thumbnails/image/monthly-sst-lanina-normal-elnino.gif).

Impacts of El Niño

The 2014-2016 El Niño was one of the strongest ENSO events on record (Figure 9.6.4). Some of the recorded global impacts of this El Niño included:

- Widespread droughts in the Philippines and many South Pacific island nations.
- Severe coral bleaching on the Great Barrier Reef in Australia.
- One of the most destructive bushfire seasons in Australia, in part due to low rainfall.
- High rainfall in the southeastern United States and parts of California, leading to flooding.
- Mild, low-precipitation winter in the New England region of the United States.

- Severe flooding in Peru and Argentina.
- Droughts in many portions of southern Africa.
- Nearly 100 million people worldwide suffered a lack of food or water from flooding and droughts.
- Peru suspended its second anchovy fishing season due to low biomass, and an anticipated 20% reduction in the yearly catch.

9.7 LANGMUIR CIRCULATION

Another sort of surface current caused by wind happens on a smaller scale when the winds consistently blow in one direction at relatively high velocities. Beneath these powerful gusts, water moves in a sequence of parallel spiraling patterns known as Langmuir cells (Figure 9.7.1 top).

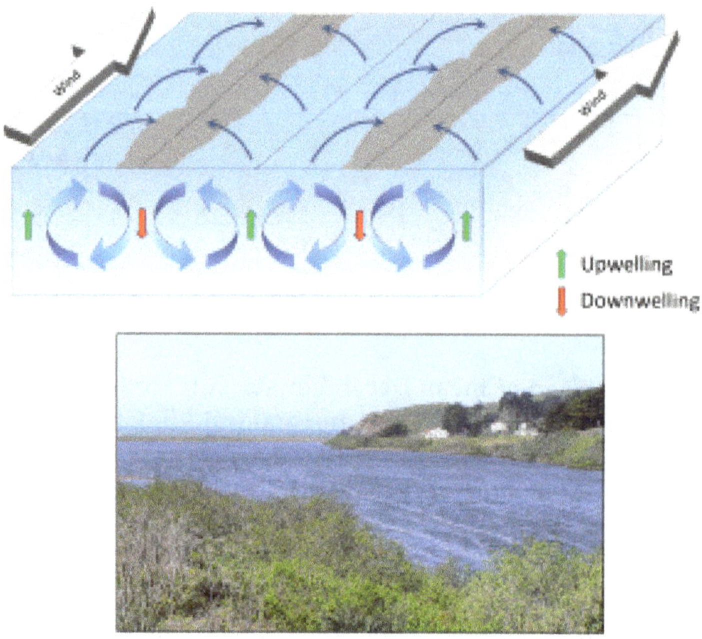

Figure 9.7.1 (Top) Langmuir cells created by strong, sustained winds (gray arrows). Dark blue arrows show the movement of surface water as a result of the cells, producing downwelling convergence zones where floating material can accumulate (gray shading), and upwelling divergence zones. (Bottom) Parallel patterns of accumulated foam and debris indicating Langmuir circulation (Top: PW; Bottom: Photo By Mmelugin (Own work) [CC BY-SA 3.0 or GFDL (http://www.gnu.org/copyleft/fdl.html)], via Wikimedia Commons).

Each cell has dimensions of several meters in width and depth, and neighboring cells rotate in opposite directions. The corkscrew motion primarily aligns with the wind direction, whereas the cells rotate approximately at a right angle to the wind direction. A divergence zone forms when the surfaces of adjacent cells rotate in opposite directions, causing an upward flow of material between the cells. A convergent zone is created when the surface water of adjacent cells rotates towards each other, resulting in a region of downwelling between the cells. The alternating regions of divergence and convergence frequently gather debris, foam, or algae that are floating on the surface and concentrate them in the convergence zones. This process results in the formation of long slicks that run parallel to the direction of the wind (Figure 9.7.1 bottom).

9.8 THERMOHALINE CIRCULATION

The surface currents we have discussed so far are ultimately driven by the wind, and since they only involve surface water they only affect about 10% of the ocean's volume. However, there are other significant ocean currents that are independent of the wind, and involve water movements in the other 90% of the ocean. These currents are driven by differences in water density.

Recall that less dense water remains at the surface, while denser water sinks. Waters of different densities tend to stratify themselves into layers, with the densest, coldest water on the bottom and warmer, less dense water on top. It is the movement of these density layers that create the deep water circulation. Since seawater density depends mainly on temperature and salinity (section 6.3), this circulation is referred to as **thermohaline circulation**.

The main processes that increase seawater density are cooling, evaporation, and ice formation. Evaporation and ice formation cause an increase in density by removing fresh water, leaving the remaining seawater with greater salinity (see section 5.3). The main processes that decrease seawater density are heating, and dilution by fresh water through precipitation, ice melting, or fresh water runoff. Note that all of these processes exert their effects at the surface, but don't necessarily affect deeper water. However, changing the density of the surface water causes it to sink or rise, and these vertical, density-driven movements create the deep ocean currents. These thermohaline currents are slow, on the order of 10-20 km per year compared with surface currents that move at several kilometers per hour.

Water masses

A water mass is a volume of seawater with a distinctive density as a result of its unique profile of temperature and salinity. As stated above, the processes that affect seawater density really only happen at the surface. Once a water mass has reached its particular temperature and salinity profile due to these surface processes, it may sink below the surface, at which point its density properties won't really change. We can therefore distinguish particular water masses by taking salinity and temperature measurements at different depths, and looking for the unique combination of these variables that give it its characteristic density. This is often carried out using temperature-salinity diagrams.

There are several well-known water masses in the ocean, particularly in the Atlantic, that are distinguished by their temperature and salinity characteristics. The densest ocean water is formed in two primary locations near the poles, where the water is very cold and highly saline as a result of ice formation. The densest deep water mass is formed in the Weddell Sea of Antarctica, and becomes the **Antarctic Bottom Water (AABW)**. Similar processes in the North Atlantic produce the **North Atlantic Deep Water (NADW)** in the Greenland Sea (Figure 9.8.1).

Figure 9.8.1 The primary sites of deep water formation; Antarctic Bottom Water is formed int he Weddell Sea, and North Atlantic Deep water is formed in the Greenland Sea (PW).

This cold, dense water sinks, and once it is removed from the surface, its temperature and salinity remain unchanged, so it keeps the same characteristics as it moves throughout the ocean as part of the thermohaline circulation. AABW sinks to the bottom in the Weddell Sea and then moves north along the bottom into the Atlantic, and east through the Southern Ocean. At the same time NADW is sinking in the Greenland Sea. This water mass is less dense than AABW and tends to form a layer above the AABW as it flows across the equator to the south (Figure 9.8.2). As the NADW moves towards the Antarctic continent, it is brought to the surface. Recall that near Antarctica there is the Antarctic divergence, where surface waters move horizontally away from each other, and are replaced by deep water upwelling (bringing nutrients to the surface and leading to high productivity; see section 7.3). Since polar water has a weak thermocline, there isn't much of a density difference preventing the deep water from reaching the surface, so some NADW rises as part of the upwelling process (Figure 9.8.2).

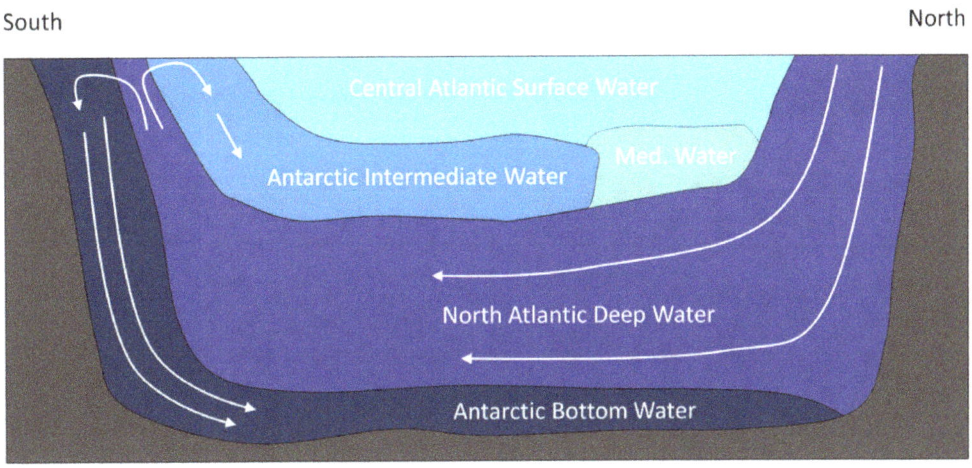

Figure 9.8.2 The major water masses of the Atlantic Ocean (PW).

As the rising NADW reaches the surface, some travels south where it will eventually contribute to the production of new AABW. The NADW that moves north encounters the Antarctic convergence, which produces downwelling. This sinking NADW becomes a new water mass; **Antarctic Intermediate Water (AAIW)**, which sinks and creates a layer between the surface water and the NADW (Figure 9.8.2). The surface water in the equatorial Atlantic, also called the **Central Atlantic Surface Water**, is very warm and low density, therefore it remains at the surface and does not contribute much to thermohaline circulation.

In the Atlantic, **Mediterranean Intermediate Water (MIW)** flows through the Straits of Gibraltar into the open ocean. This water is warm and salty from the warm temperatures and high evaporation characteristic of the Mediterranean Sea, so it is denser than the normal surface water and forms a layer about 1-1.5 km deep. Eventually this water will move north to the Greenland Sea, where it will be cooled and will sink, becoming the dense NADW.

T-S Diagrams

A temperature-salinity (T-S) diagram is used to examine how temperature, salinity, and density change with depth, and to identify the vertical structure of the water column, including the water masses it contains. Water temperature is on the y-axis, and salinity appears on the x-axis. Often, instead of the actual temperature of the water, oceanographers plot **potential temperature,** which is the temperature the water would achieve if it was brought to the surface and did not get any additional heat through compression at depth. A T-S diagram shows lines of equal density, or **isopycnals**, for various combinations of temperature and salinity (Figure 9.8.3). You can then plot the values of temperature and salinity on the diagram, and use their point of intersection to calculate the density of the water. In the example in Figure 9.8.3 a temperature of about 11° C and a salinity of 34.6 PSU results in a density of 1.0265 g/cm^3.

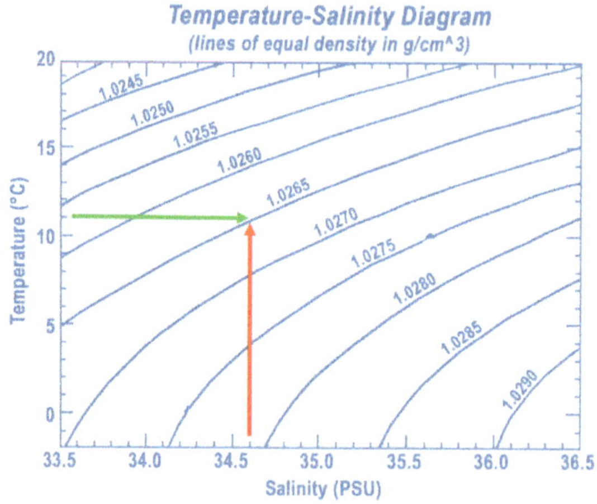

Figure 9.8.3 Using a T-S diagram to determine density. A temperature of about 11° C (green arrow) and a salinity of 34.6 PSU (red arrow) results in a density of 1.0265 g/cm^3.

Since the range of densities in the ocean is rather small, often the density value is shortened and is expressed as sigma-t or σ_t. Sigma-t is calculated as: (density – 1) x 1000. So it essentially just looks at the last three decimal places of the density value. Thus a density of 1.0275 g/cm^3 would have a σ_t of 27.5.

T-S diagrams can be used to identify water masses. Since each major water mass has its own characteristic range of temperatures and salinities, a deep water sample that falls into that range can presumably have come from that water mass. Figure 9.8.4 shows the typical range of temperature and salinity for the major Atlantic water masses.

To investigate water masses, oceanographers can take a series of temperature and salinity measurements over a range of depths at a particular location. If the water column was highly stratified and there was no mixing between or within the layers, as the probe was lowered your would get a series of constant temperature and salinity readings as you moved through the first water mass, followed by a sudden jump to another set of different but constant readings as you moved through the next water mass. Plotting temperature vs. salinity on a T-S diagram would result in a distinct and independent point for each water mass. However, in reality, the water masses will show some mixing within and between layers. So as the probes are lowered, they will encounter water that shows traits intermediate between the two points. Therefore, with increasing depth, the points on the T-S diagram will gradually move from one point to the other, creating a line connecting the two points, illustrating mixing between those two water masses.

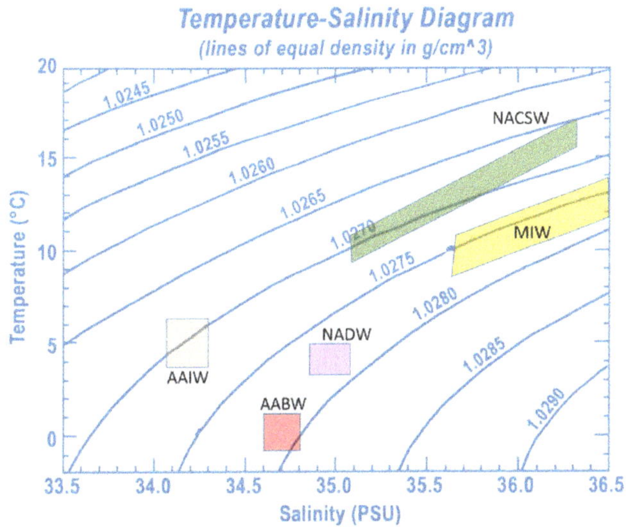

Figure 9.8.4 Characteristic ranges of temperature and salinity for the major Atlantic water masses; North Atlantic Central Surface Water (NACSW), Mediterranean Intermediate Water (MIW), Antarctic Intermediate Water (AAIW), North Atlantic Deep Water (NADW), and Antarctic Bottom Water (AABW).

In the example in Figure 9.8.5, NACSW is present at the surface (0 m depth), and between 0 and about 800 m there is a transition from NACSE into AAIW. Between about 800-2100 m there is a transition from AAIW into the NADW layer just beyond 2000 m. AABW is the deepest water mass, at depths of about 4000 m. The transition between NADW and AABW occurs between about 2100-4000 m.

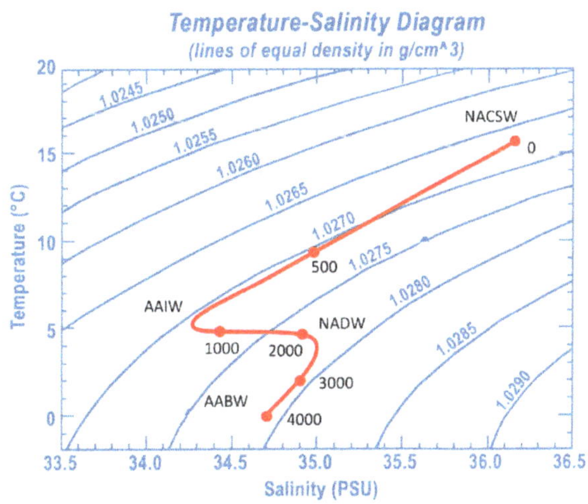

Figure 9.8.5 Hypothetical T-S diagram for the North Atlantic. Points represent readings taken at the corresponding depths (m). Moving from the surface to the bottom results in water of increasing density, passing through distinct water masses.

Notice that as the recordings get deeper in Figure 9.8.5, the density is always increasing (i.e. moving towards the bottom right corner). This is because the densest water should be located at the bottom, with the other layers stratified according to their density, otherwise the water column would be unstable.

The "Ocean Conveyor Belt"

The bottom water from the Weddell Sea and Greenland Sea does not just circulate through the Atlantic. NADW moves south through the western Atlantic before meeting the AABW north of the Weddell Sea. Together these water masses move eastwards into the Indian and Pacific Oceans. By this time the NADW and AABW have started mixing, to create what is called **Common Water**. The deep Common Water moves northwards into the Pacific and Indian Oceans and gradually mixes with the warmer water, causing it to eventually rise to the surface. As surface water, it makes its way back to the North Atlantic through the surface currents of the Pacific and Indian Oceans. Once back in the North Atlantic, it cools and once again forms NADW, starting the process anew. This cycle of rising and sinking water transporting water between the surface and deep circulation has been referred to as the global oceanic "conveyor belt", and may take about 1000-2000 years to complete (Figure 9.8.6).

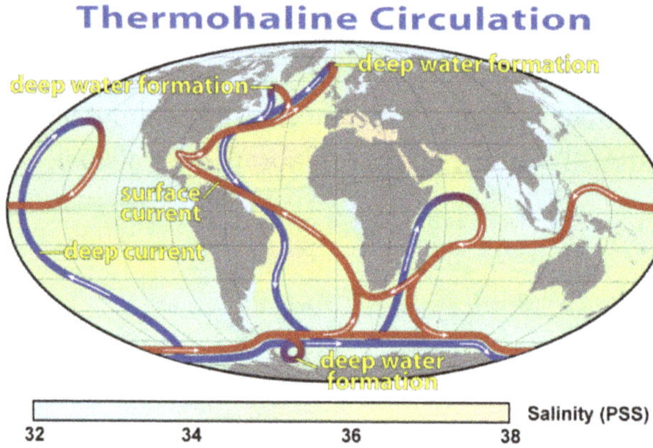

Figure 9.8.6 The global ocean "conveyor belt." Cold, dense water sinks in the Greenland and Weddell Seas and circulates over the seafloor into the Indian and Pacific Oceans (blue paths). Eventually the water rises to the surface, and returns to the site of bottom water formation via surface currents (red paths), to start the cycle again (By Robert Simmon, NASA. Minor modifications by Robert A. Rohde also released to the public domain (NASA Earth Observatory) [Public domain], via Wikimedia Commons).

The worldwide circulation pattern has significant ramifications for Earth's environment. Firstly, it is crucial for the circulation of heat worldwide, facilitating the movement of warm water towards the poles and cold water towards the tropics, thereby maintaining temperature stability in both regions.

The conveyor belt facilitates the transportation of oxygen to habitats located in the deep sea. The cold surface water, saturated with oxygen, transformed into deep water and transported the oxygen to lower depths. The thermohaline circulation transports oxygen-rich deep water across the oceans, providing a source of oxygen for deep water life. The oxygen content in bottom water of the Atlantic is relatively high due to its retention of a significant portion of its original oxygen. However, as it moves across the seafloor, the oxygen is depleted. Consequently, deep water in the Pacific Ocean contains considerably less oxygen compared to deep Atlantic water, while Indian Ocean water falls somewhere in between. Simultaneously, profound depths will amass nutrients through the sinking and decomposition of organic materials. The nutritional content of Atlantic bottom water is low due to limited time for accumulation and the initial nutrient-poor nature of the surface water. When the bottom water reaches the Indian Ocean and then the Pacific, it has been gathering nutrients that sink for hundreds of years. As a result, the Pacific contains higher concentrations of deep-sea nutrients compared to the Atlantic. Consequently, we can utilize the proportions of oxygen to nutrients in the deep water to determine the comparative age of a water mass, specifically the duration since it descended from the surface. Bottom water that is younger should have a high oxygen content and a low nutrient content, whereas older bottom water would be expected to have the reverse characteristics.

Climate change has the potential to dramatically disrupt thermohaline circulation, which in turn can have a significant impact on the ocean conveyor belt. Rising temperatures, especially in the Arctic region, may result in ongoing melting of the polar ice caps, contributing a significant volume of freshwater to the polar surface water. The introduction of new freshwater might result in the formation of a surface layer of water with low density and salinity. This layer would no longer sink, which would disrupt the deep circulation conveyor belt and hinder the transportation of oxygen and nutrients to the populations at the bottom. The subsidence of saltwater in the Greenland Sea also contributes to the propulsion of the Gulf Stream; when water subsides, it induces the northward movement of more surface water in the Gulf Stream. If the process of polar water sinking ceases, it could lead to a weakening of the Gulf Stream, resulting in a reduction in heat transportation towards the poles and ultimately causing a cooling effect on the northern climate. Paradoxically, global warming may result in colder weather in Europe, causing the freezing of often ice-free ports and towns, as a consequence of the Gulf Stream's warming impacts. Recent research indicates that the intensity of the Gulf Stream is diminishing, most likely as a result of the accelerated melting of Arctic ice.

10 Chapter | WAVES

INTRODUCTION: AN OVERVIEW

Waves exhibit various forms and dimensions; while a 100-foot wave may be highly desirable for a surfer, it might be extremely distressing for a ship captain. What is the record for the greatest wave ever measured? Is it 50 feet? Is it 100 feet? Nowhere near. The record pertains to a wave generated in Lituya Bay, Alaska, on July 9, 1958 (Figure 10.1). On that particular day, a seismic event of magnitude 7.8 triggered a substantial rockslide that descended a mountainside and entered the source of the bay.

Figure 10.1 A view of Lituya Bay taken a few weeks after the 1958 megatsunami. The rockslide occurred in the mountains at the head of the bay, producing the wave that them moved through the bay towards the sea (D.J. Miller, United States Geological Survey, [Public domain], via Wikimedia Commons).

The rockslide generated a tsunami-like wave that reached an elevation of 1722 ft (525 m) above sea level, causing the destruction of flora in its path. Subsequently, the wave advanced through the constricted harbor towards the open ocean, causing significant damage to several fishing vessels in its path. Remarkably, a father and son aboard a fishing vessel were lifted over the tree line by the wave, and managed to survive in order to recount the tale. This megatsunami is the greatest wave ever recorded with reliable evidence. The waves we will examine in this chapter may not be as spectacular, but it is nevertheless crucial to understand their formation, propagation, and the consequences of their interaction with the coast.

10.1 WAVE BASICS

- Waves often originate from a disturbance and the energy from this disturbance is transmitted in the form of waves. While we are well acquainted with the waves that crash on the shore or cause a boat to sway in the water, there are numerous more varieties of waves that hold significance in the field of oceanography:
- **Internal waves** form at the boundaries of water masses of different densities (i.e. at a pycnocline), and propagate at depth. These generally move more slowly than surface waves, and can be much larger, with heights exceeding 100 m. However, the height of the deep wave would be unnoticeable at the surface.
- **Tidal waves** are due to the movement of the tides. What we think of as tides are basically enormously long waves with a wavelength that may span half the globe (see section 11.1). Tidal waves are not related to tsunamis, so don't confuse the two.
- **Tsunamis** are large waves created as a result of earthquakes or other seismic disturbances. They are also called seismic sea waves (section 10.4).
- **Splash waves** are formed when something falls into the ocean and creates a splash. The giant wave in Lituya Bay that was described in the introduction to this chapter was a splash wave.
- **Atmospheric waves** form in the sky at the boundary between air masses of different densities. These often create ripple effects in the clouds (Figure 10.1.1).

There are several components to a basic wave (Figure 10.1.2):

- **Still water level:** where the water surface would be if there were no waves present and the sea was completely calm.
- **Crest:** the highest point of the wave.
- **Trough:** the lowest point of the wave.
- **Wave height:** the distance between the crest and the trough.

- **Wavelength:** the distance between two identical points on successive waves, for example crest to crest, or trough to trough.
- **Wave steepness:** the ratio of wave height to length (H/L). If this ratio exceeds 1/7 (i.e. height exceeds 1/7 of the wavelength) the wave gets too steep, and will break.

Figure 10.1.1 Wake patterns in cloud cover over Possession Island, East Island, Ile aux Cochons, Ile de Pingouins. The ripple pattern is a result of internal waves in the atmosphere (NASA [Public domain], via Wikimedia Commons).

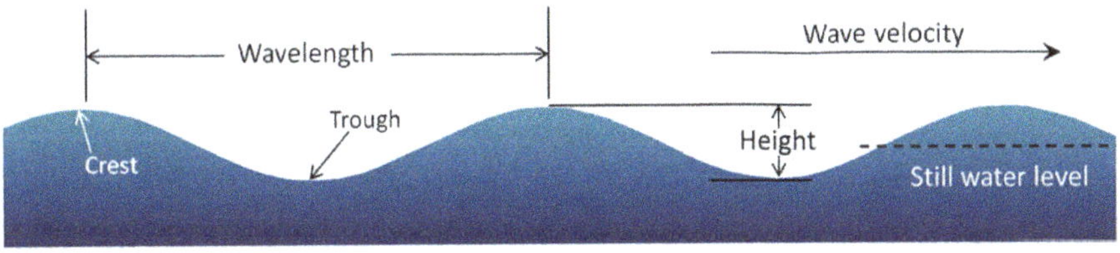

Figure 10.1.2 Components of a basic wave (Modified by PW from Steven Earle "Physical Geology").

There are also a number of terms used to describe wave motion:

- **Period:** the time it takes for two successive crests to pass a given point.
- **Frequency:** the number of waves passing a point in a given amount of time, usually expressed as waves per second. This is the inverse of the period.
- **Speed:** how fast the wave travels, or the distance traveled per unit of time. This is also called celerity (c), where

$$c = \text{wavelength} \times \text{frequency}$$

Consequently, waves with longer wavelengths have a higher speed.

While waves have the ability to travel long distances, the water itself exhibits minimal horizontal motion. It is important to note that it is the energy of the wave that is being conveyed, rather than the water itself. Conversely, the water particles exhibit circular motion, where the diameter of the circle is equal to the height of the wave (Figure 10.1.3). The orbital motion of water waves arises from the presence of both longitudinal (horizontal) and transverse (vertical) wave components, resulting in a circular motion. As a wave propagates, water undergoes forward and upward motion over the wave crests, followed by downward and backward motion into the troughs, resulting in minimal horizontal displacement. This is apparent if you have ever observed an object, such as a seabird, resting on the surface. The bird oscillates vertically as the waves pass under it; it does not get horizontally displaced by a single wave crest.

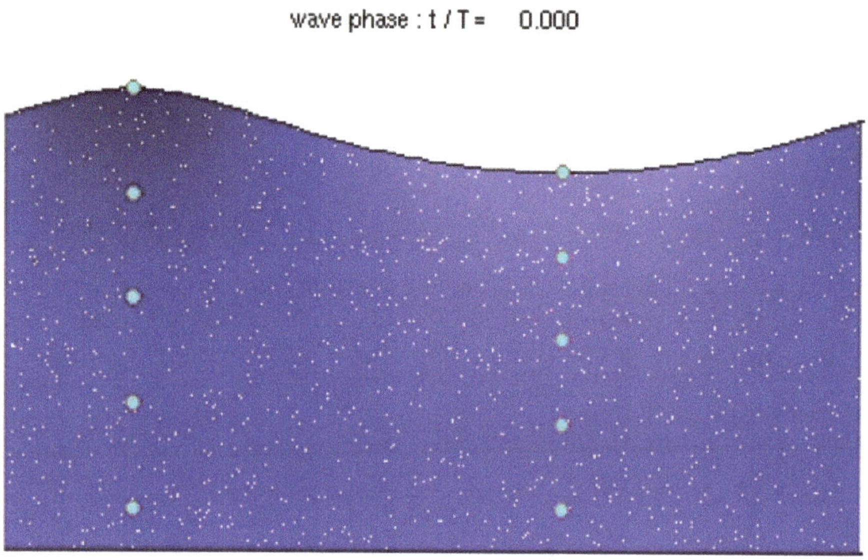

Figure 10.1.3 Animation showing the orbital motion of particles in a surface wave (By Kraaiennest (Own work) [GFDL (http://www.gnu.org/copyleft/fdl.html) or CC BY-SA 4.0], via Wikimedia Commons).

The circular orbital motion diminishes with increasing depth due to the decreased influence of the wave on deeper water, resulting in a reduction in the diameter of the circles. Ultimately, at a certain depth, the water reaches a point where there is no longer any circular motion and the surface wave action has no impact on it. The depth referred to as the wave base is precisely half the wavelength, as shown in Figure 10.1.4. Due to the relatively short wavelengths of most ocean waves, which are typically less than a few hundred meters, the majority of the deeper ocean remains undisturbed by surface waves. Consequently, even during severe storms, marine life or submarines can evade the impact of large waves by diving below the level at which the waves propagate.

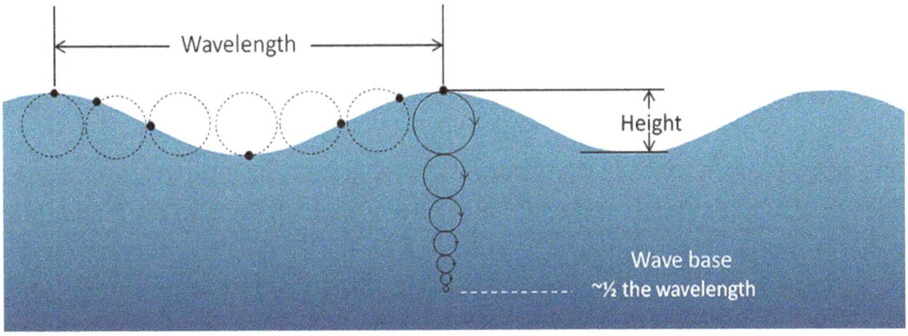

Figure 10.1.4 Orbital motion of water within a wave, extending down to the wave base at a depth of half of the wavelength (Modified by PW from Steven Earle, "Physical Geology").

When the water below a wave is deeper than the wave base (deeper than half of the wavelength), those waves are called **deep water waves.** Most open ocean waves are deep water waves. Since the water is deeper than the wave base, deep water waves experience no interference from the bottom, so their speed only depends on the wavelength:

$$\text{speed (m/s)} = \sqrt{\frac{gL}{2\pi}}$$

where g is gravity and L is wavelength in meters. Since g and π are constants, this can be simplified to:

$$\text{speed (m/s)} = 1.25\sqrt{L}$$

Shallow water waves occur when the depth is less than 1/20 of the wavelength. In these cases, the wave is said to "touch bottom" because the depth is shallower than the wave base so the orbital motion is affected by the seafloor. Due to the shallow depth, the orbits are flattened, and eventually the water movement becomes horizontal rather than circular just above the bottom. The speed of shallow water waves depends only on the depth:

$$\text{speed (m/s)} = \sqrt{gd}$$

where g is gravity and d is depth in meters. This can be simplified to:

$$\text{speed (m/s)} = 3.13\sqrt{d}$$

Intermediate or **transitional waves** are found in depths between ½ and 1/20 of the wavelength. Their behavior is a bit more complex, as their speed is influenced by both wavelength and depth. The speed of an intermediate wave is calculated as:

$$\text{speed (m/s)} = \sqrt{\frac{gL}{2\pi}\tanh(2\pi\frac{d}{L})}$$

which contains both depth and wavelength variables.

10.2 WAVES AT SEA

The majority of ocean waves are created by the force of wind. Capillary waves, also known as ripples, are small disturbances that form on the surface of water due to the blowing of wind (Figure 10.2.1). Capillary waves exhibit a crest that is rounded in shape and a trough that has a V-shaped structure. Additionally, these waves have wavelengths that are shorter than 1.7 cm. The presence of these minor undulations provides the wind with a surface to hold onto, resulting in the formation of more substantial waves as the wind intensity escalates. Once the wavelength surpasses 1.7 cm, the wave transforms from a capillary wave into a wind wave. When waves are generated, they encounter a restorative force that strives to bring the water back to its tranquil and balanced state. Surface tension acts as the restoring force for small capillary waves, whereas gravity serves as the restoring force for larger wind-generated waves.

Figure 10.2.1 Small capillary waves or ripples caused by winds blowing over the surface of calm water (By Blue Elf (Own work) [GFDL (http://www.gnu.org/copyleft/fdl.html) or CC BY-SA 3.0], via Wikimedia Commons).

As the energy of the wind increases, so does the size, length and speed of the resulting waves. There are three important factors determining how much energy is transferred from wind to waves, and thus how large the waves will get:

- Wind **speed.**
- The **duration** of the wind, or how long the wind blows continuously over the water.
- The distance over which the wind blows across the water in the same direction, also known as the **fetch.**

Increasing any of these factors increases the energy of wind waves, and therefore their size and speed. But there is an upper limit to how large wind-generated waves can get. As wind energy increases, the waves receive more energy and they get both larger and steeper (recall from section 10.1 that wave steepness = height/wavelength). When the wave height exceeds 1/7 of the wavelength, the wave becomes unstable and collapses, forming whitecaps.

The ocean surface represents an irregular mixture of hundreds of waves of different speeds and sizes, all coming from different directions and interacting with each other. A histogram of wave heights within this mixture reveals a bell-shaped curve (Figure 10.2.2). In addition to basic statistics such as mode (most probable), median and mean wave height, wave heights are also reported in other ways. Marine weather forecasts and ship and buoy data often report **significant wave height (H_s)**, which is the mean height of the largest one-third of the waves. Mean wave height is approximately equal to two-thirds of the significant wave height. Finally, there is the minimum height of the highest 10% of waves (the 90th percentile of wave heights), often expressed as $H_{1/10}$.

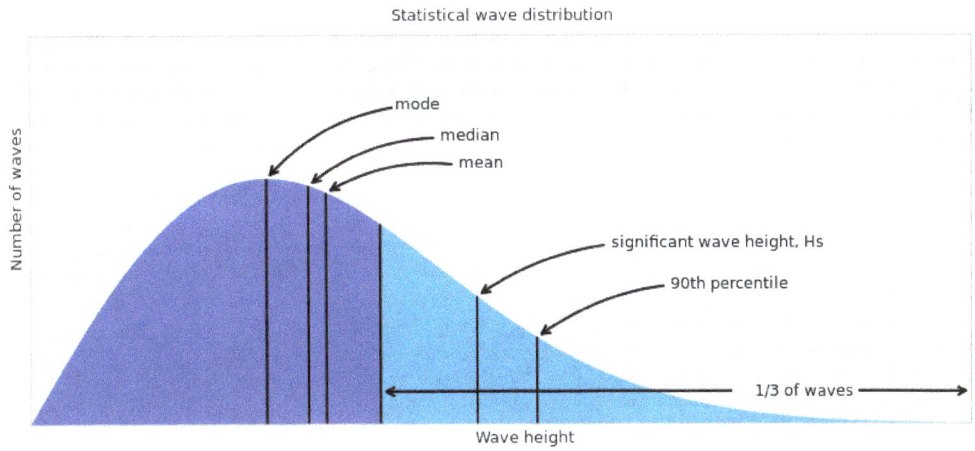

Figure 10.2.2 Histogram of typical wave height distribution at sea, showing common statistical measurements (NOAA, Public Domain via Wikimedia Commons).

Under strong wind conditions, the ocean surface becomes a chaotic mixture of choppy, whitecapped wind-generated waves. The term **sea state** describes the size and extent of the wind-generated waves in a particular area. When the waves are at their maximum size for the existing wind speed, duration, and fetch, it is referred to as a fully developed sea. The sea state is often reported on the **Beaufort scale**, ranging from 0-12, where 0 means calm, windless and waveless conditions, while Beaufort 12 is a hurricane.

The Beaufort Scale

The Beaufort scale is used to describe the wind and sea state conditions on the ocean. It is an observational scale based on the judgement of the observer, rather than one dictated by accurate measurements of wave height. Beaufort 0 represents calm, flat conditions, while Beaufort 12 represents a hurricane.

BEAUFORT FORCE 0
WIND SPEED: LESS THAN 1 KNOT

SEA: SEA LIKE A MIRROR

BEAUFORT FORCE 1
WIND SPEED: 1-3 KNOTS

SEA: WAVE HEIGHT .1M (.25FT), RIPPLES WITH THE APPEARANCE OF SCALES, BUT WITHOUT FOAM CRESTS

BEAUFORT FORCE 2
WIND SPEED: 4-6 KNOTS

SEA: WAVE HEIGHT .2-.3M (.5-1FT), SMALL WAVELETS, CRESTS HAVE A GLASSY APPEARANCE AND DO NOT BREAK

BEAUFORT FORCE 3
WIND SPEED: 7-10 KNOTS

SEA: WAVE HEIGHT .6-1M (2-3FT), LARGE WAVELETS, CRESTS BEGIN TO BREAK. ANY FOAM HAS GLASSY APPEARANCE. SCATTERED WHITECAPS

BEAUFORT FORCE 4
WIND SPEED: 11-16 KNOTS

SEA: WAVE HEIGHT 1-1.5M (3.5-5FT), SMALL WAVES BECOMING LONGER. FAIRLY FREQUENT WHITE HORSES

BEAUFORT FORCE 5
WIND SPEED: 17-21 KNOTS

SEA: WAVE HEIGHT 2-2.5M (6-8FT), MODERATE WAVES TAKING MORE PRONOUNCED LONG FORM, MANY WHITE HORSES, CHANCE OF SOME SPRAY

BEAUFORT FORCE 6
WIND SPEED: 22-27 KNOTS

SEA: WAVE HEIGHT 3-4M (9.5-13 FT), LARGER WAVES BEGIN TO FORM, SPRAY IS PRESENT, WHITE FOAM CRESTS ARE EVERYWHERE

BEAUFORT FORCE 7
WIND SPEED: 28-33 KNOTS

SEA: WAVE HEIGHT 4-5.5M (13.5-19 FT), SEA HEAPS UP, WHITE FOAM FROM BREAKING WAVES BEGINS TO BE BLOWN IN STREAKS ALONG THE WIND DIRECTION

BEAUFORT FORCE 8
WIND SPEED: 34-40 KNOTS

SEA: WAVE HEIGHT 5.5-7.5M (18-25FT), MODERATELY HIGH WAVES OF GREATER LENGTH, EDGES OF CREST BEGIN TO BREAK INTO THE SPINDRIFT, FOAM BLOWN IN WELL MARKED STREAKS ALONG WIND DIRECTION.

BEAUFORT FORCE 9
WIND SPEED: 41-47 KNOTS

SEA: WAVE HEIGHT 7-10M (23-32FT), HIGH WAVES, DENSE STREAKS OF FOAM ALONG DIRECTION OF THE WIND, WAVE CRESTS BEGIN TO TOPPLE, TUMBLE, AND ROLL OVER. SPRAY MAY AFFECT VISIBILITY.

BEAUFORT FORCE 10
WIND SPEED: 48-55 KNOTS

SEA: WAVE HEIGHT 9-12.5M (29-41FT), VERY HIGH WAVES WITH LONG OVERHANGING CRESTS. THE RESULTING FOAM, IN GREAT PATCHES, IS BLOWN IN DENSE WHITE STREAKS ALONG WIND DIRECTION. ON THE WHOLE, SEA SURFACE TAKES A WHITE APPEARANCE. TUMBLING OF THE SEA IS HEAVY AND SHOCK-LIKE, VISIBILITY AFFECTED.

BEAUFORT FORCE 11
WIND SPEED: 56-63 KNOTS

SEA: WAVE HEIGHT 11.5-16M (37-52FT), EXCEPTIONALLY HIGH WAVES, SMALL-MEDIUM SIZED SHIPS MAY BE LOST TO VIEW BEHIND THE WAVES. SEA COMPLETELY COVERED WITH LONG WHITE PATCHES OF FOAM LYING ALONG WIND DIRECTION. EVERYWHERE, THE EDGES OF WAVE CRESTS ARE BLOWN INTO FROTH.

(Images by United States National Weather Service (http://www.crh.noaa.gov/mkx/marinefcst.php) [Public domain], via Wikimedia Commons).

A mature oceanic environment frequently emerges during turbulent weather conditions, characterized by strong winds that generate a disordered and unpredictable arrangement of waves and foamy crests of different magnitudes. The waves will radiate outward from the storm's center, propelled by the powerful winds. Nevertheless, once the storm diminishes and the winds decrease in intensity, these chaotic ocean waves will gradually organize themselves into more structured formations. It is important to remember that open ocean waves are typically deep water waves, and their velocity is determined by their wavelength (section 10.1). As the waves disperse from the center of the storm, they arrange themselves according to their speed, with waves of longer wavelength flowing at a higher velocity than waves of shorter wavelength. Consequently, over time, all the waves in a specific region will move with an identical wavelength, resulting in the formation of uniform, extended waves known as swell (Figure 10.2.3). Swell refers to the gradual oscillation or rocking movement felt on a boat, or the consistent arrival of waves on the coast. Swell waves have the ability to travel over great distances without significant energy loss. As a result, we can see the arrival of massive swells at the coastline, even in the absence of local wind. These waves are generated by storms that occur far away from the coast and transform into swells as they journey towards the shoreline.

Figure 10.2.3 Ocean swell, the regular pattern of waves of equal wavelength (Phillip Capper [CC BY 2.0], via Wikimedia Commons).

Because swell travels such long distances, eventually swells coming from different directions will run into each other, and when they do they create interference patterns. The interference pattern is created by adding the features of the waves together, and the type of interference that is created depends on how the waves interact with each other (Figure 10.2.4). **Constructive interference** occurs when the two waves are completely in phase; the crest of one wave lines up exactly with the crest of the other wave, as do the troughs of the two waves. Adding the two crest together creates a crest that is higher than in either of the source waves, and adding the troughs creates a deeper trough than in the original waves. The result of constructive interference is therefore to create waves that are larger than the original source waves. In **destructive interference**, the waves interact completely out of phase, where the crest of one wave aligns with the trough of the other wave. In this case, the crest and the trough work to cancel each other out, creating a wave that is smaller than either of the source waves. In reality, it is rare to find perfect constructive or destructive interference as displayed in Figure 10.2.4. Most interference by swells at sea is **mixed interference**, which contains a mix of both constructive and destructive interference. The interacting swells do not have the same wavelength, so some points show constructive interference, and some points show destructive interference, to varying degrees. This results in an irregular pattern of both small and large waves, called **surf beat**.

It is important to point out that these interference patterns are only temporary disturbances, and do not affect the properties of the source waves. Moving swells interact and create interference where they meet, but each wave continues on unaffected after the swells pass each other.

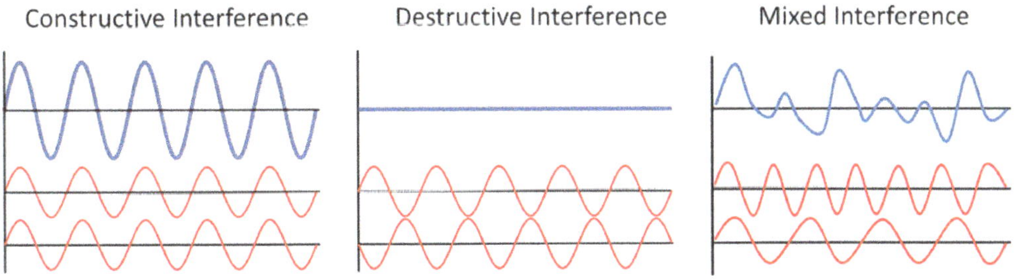

Figure 10.2.4 Wave interference patterns. In constructive interference the source waves (red) are completely in phase, and when added together produce waves that are larger than the original waves (blue). In destructive interference the source waves are out of phase, so they cancel each other out and produce waves that are smaller than the originals. In mixed interference, constructive and destructive interference occur at various point, creating an irregular wave pattern. (Modified by PW from original version: Haade; vectorization: Wjh31, Quibik (Vecorized from File:Interference of two waves.png) [CC BY-SA 3.0 or GFDL (http://www.gnu.org/copyleft/fdl.html)], via Wikimedia Commons).

Figure 10.2.5 A rogue wave in the Bay of Biscay, off of the French coast, ca. 1940 (NOAA, [Public domain], via Wikimedia Commons).

About half of the waves in the open sea are less than 2 m high, and only 10-15% exceed 6 m. But the ocean can produce some extremely large waves. The largest wind wave reliably measured at sea occurred in the Pacific Ocean in 1935, and was measured by the navy tanker the USS Ramapo. Its crew measured a wave of 34 m or about 112 ft high! Occasionally constructive interference will produce waves that are exceptionally large, even when all of the surrounding waves are of normal height. These random, large waves are called **rogue waves** (Figure 10.2.5). A rogue wave is usually defined as a wave that is at least twice the size of the significant wave height, which is the average height of the highest one-third of waves in the region. It is not uncommon for rogue waves to reach heights of 20 m or more.

Rogue waves are particularly common off of the southeast coast of South Africa, a region referred to as the "wild coast." Here, Antarctic storm waves move north into the oncoming Agulhas Current, and the wave energy gets focused over a narrow area, leading to constructive interference. This area may be responsible for sinking more ships than anywhere else on Earth. On average about 100 ships are lost every year across the globe, and many of these losses are probably due to rogue waves.

Waves in the Southern Ocean are generally fairly large (the red areas in Figure 10.2.6) because of the strong winds and the lack of landmasses, which provide the winds with a very long fetch, allowing them to blow unimpeded over the ocean for very long distances. These latitudes have been termed the "Roaring Forties", "Furious Fifties", and "Screaming Sixties" due to the high winds.

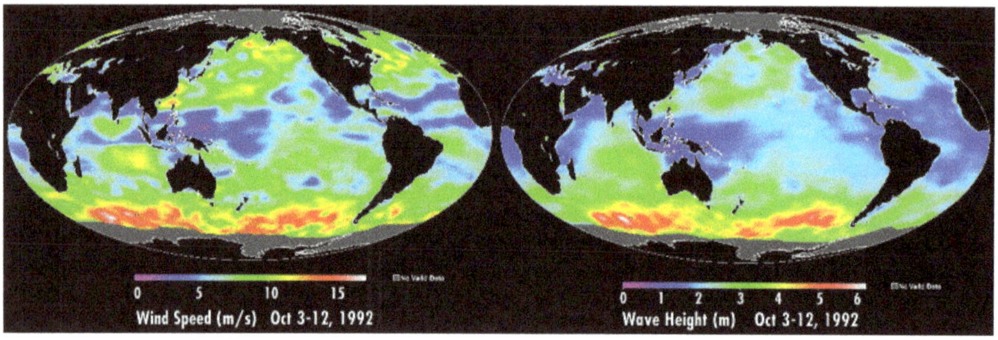

Figure 10.2.6 Wind speed and wave height data for a 9-day period in 1992. The Southern Ocean is notorious for its high winds and large waves (NASA, Public Domain via Wikimedia Commons).

10.3 WAVES ON THE SHORE

The majority of the waves covered in the preceding section pertain to deep water waves occurring in the expanse of the ocean. However, what occurs when these waves approach the coastline and come across shallow water? It is important to note that the speed of a wave in deep water is determined by its wavelength, whereas in shallow water, the wave speed is influenced by the depth (section 10.1). As waves reach the beach, they will reach the ocean floor at a depth that is half of their wavelength. This means that when the water depth matches the depth of the wave base, the wave will strike the bottom (Figure 10.3.1). At this juncture, their behavior will start to be influenced by the underlying factors.

Friction between the wave and the bottom surface decelerates the wave upon contact. As a wave decelerates, the wave immediately following it accelerates, resulting in a reduction in the wavelength. Nevertheless, the wave retains its whole energy, resulting in a rise in wave height as the wavelength decreases. Ultimately, when the wave height surpasses one-seventh of the wavelength, the wave becomes unstable and transforms into a breaker. Frequently, breakers will have a tendency to curl forward during the breaking process. This occurs because the lower component of the wave decelerates prior to the upper part, as it is the initial section to come into contact with the ocean floor. In Figure 10.3.1, the highest point of the wave, known as the crest, moves faster than the rest of the wave. However, it lacks the necessary water underneath it to provide support.

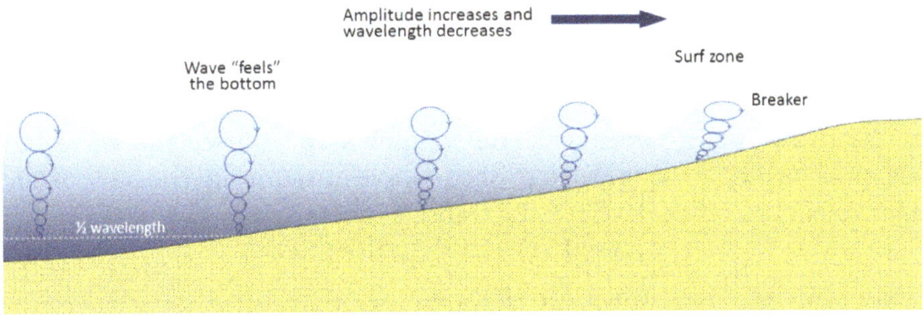

Figure 10.3.1 As waves approach shore they "touch bottom" when the depth equals half of the wavelength, and the wave begins to slow down. As is slows, the wavelength decreases and the wave height increases, until the wave breaks (Steven Earle "Physical Geology").

There are three main types of breakers: spilling, plunging, and surging. These are related to the steepness of the bottom, and how quickly the wave will slow down and its energy will get dissipated.

- **Spilling** breakers form on gently sloping or flatter beaches, where the energy of the wave is dissipated gradually. The wave slowly increases in height, then slowly collapses on itself (Figure 10.3.2). For surfers, these waves provide a longer ride, but they are less exciting.

Figure 10.3.2 A spilling breaker. The gentle slope of the bottom causes the wave height to slowly increase until the wave collapses on itself (left: JR, right: James St. John, [CC-BY-2.0], https://www.flickr.com/photos/jsjgeology/23769708334).

- **Plunging** breakers form on more steeply-sloped shores, where there is a sudden slowing of the wave and the wave gets higher very quickly. The crest outruns the rest of the wave, curls forwards and breaks with a sudden loss of energy (Figure 10.3.3). These are the "pipeline" waves that surfers seek out.

Figure 10.3.3 A plunging breaker. The steeper slope causes the wave height to increase more rapidly, with the crest of the wave outrunning the base of the wave, causing it to curl as it breaks (left: JR, right: Andrew Schmidt, Public Domain [CC-0], publicdomainpictures.net).

- **Surging** breakers form on the steepest shorelines. The wave energy is compressed very suddenly right at the shoreline, and the wave breaks right onto the beach (Figure 10.3.4). These waves give too short (and potentially painful) a ride for surfers to enjoy.

Figure 10.3.4 A surging breaker. The very steep slope causes the wave height to increase suddenly and break right on the beach (left: JR, right: Tewy, [CC-BY-SA-3.0], via Wikimedia Commons).

Wave Refraction

Ocean swells can originate from any location in the ocean, allowing them to approach a shore from virtually any direction. However, if you have ever observed the shoreline, you may have noticed that the waves tend to approach the coast in a rather parallel manner. This phenomenon is a result of wave refraction. When a wave front approaches the shore at an angle, the part of the wave front that is closest to the shore will make contact with the bottom before the rest of the wave. This will result in the deceleration of the shallower portion of the wave before the remaining portion of the wave, which is still in deeper water, continues to move at its original velocity. As the wave front gradually enters shallower water and decelerates, it undergoes refraction, causing the waves to align themselves almost parallel to the shoreline (they are bent towards the area of slower speed). In section 13.2, it will be demonstrated that the waves' lack of perfect parallelism to the beach results in the formation of longshore currents and longshore transport that flow in the same direction as the shoreline.

Refraction can account for the phenomenon of waves being amplified on points and headlands, while being diminished in bays. The wave front will make contact with the ocean floor around the point before reaching the bay. Once again, the portion of the wave front that is closer to the surface will experience a decrease in speed, resulting in the bending of the rest of the wave front towards the region with slower speed (the point). Currently, all of the initial wave energy is focused in a very compact region near the point, resulting in the formation of massive, powerful waves (Figure 10.3.5). Within the bay, the phenomenon of refraction has resulted in the divergence of wave fronts, resulting in the dispersion of wave energy. Consequently, this has led to the presence of tranquil water and diminished wave height. The substantial swells of a "point break" are optimal for surfing, whereas the sea is more tranquil in a bay, which is a suitable location for boat launching. The disparity in wave energy accounts for the occurrence of erosion on points and the deposition of sand and sediments in bays. (see section 13.3).

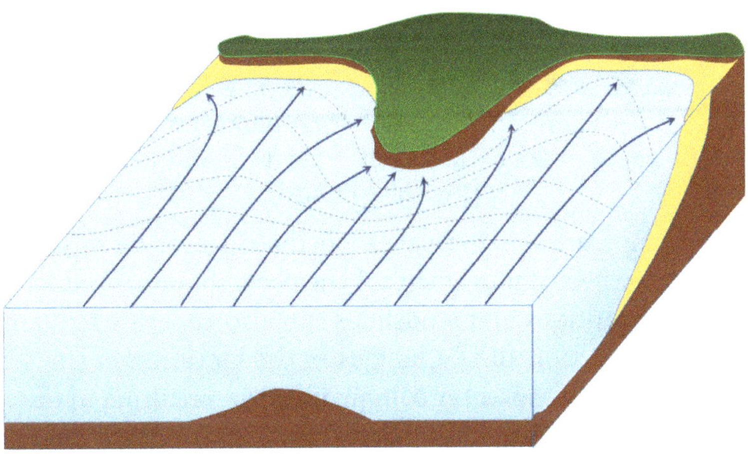

Figure 10.3.5 Waves approaching shore (blue lines) touch bottom sooner off of points and are refracted towards the points, concentrating their wave energy. Wave energy is spread out in bays, causing smaller waves. Dotted lines represent the bottom contours (PW).

10.4 TSUNAMIS

Tsunamis hold significant prominence in popular culture, although there exist some myths regarding these immense waves. Tsunamis are not related to the tides, hence it is incorrect to call them "tidal waves." Tidal waves, as described in section 11.1, are distinct from tsunamis. Furthermore, the depiction of a colossal, curving wave surpassing the height of buildings and obliterating cities in science fiction films is entirely fictional. Tsunamis, as explained below, do not exhibit such behavior.

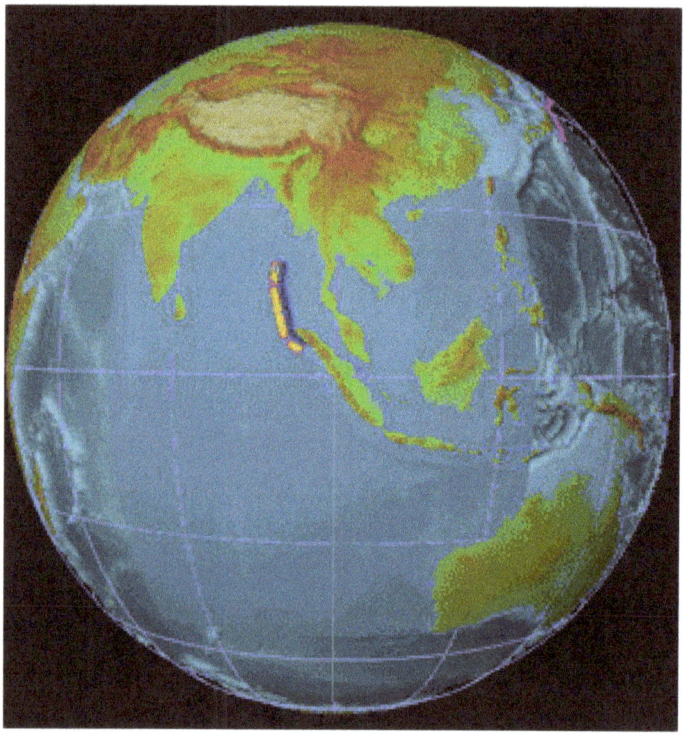

Figure 10.4.1 Animation of the spread of tsunamis created during the 2004 Indonesia earthquake (NOAA Center for Tsunami Research (NCTR) [Public domain]).

Tsunamis are immense oceanic waves mostly triggered by seismic events, such as the upward or downward movement of the ocean floor caused by earthquakes. However, volcanic eruptions and landslides can also generate tsunamis known as splash waves (see to section 10.1). Changes in the elevation of the seafloor cause corresponding changes in the water column above it, resulting in the formation of waves. Tsunamis are solely caused by vertical earthquake shocks, not by horizontal motions. The vertical changes of the seafloor typically have a maximum height of less than 10 meters, which means that the corresponding wave at sea will be of equal or lesser height. Although tsunamis may have a relatively low height at their starting point, they possess remarkably lengthy wavelengths spanning between

100 and 200 kilometers. Due to their extended wavelength, these waves exhibit characteristics similar to those of shallow water waves across the whole ocean. The ocean's depth is consistently less than half of their wavelength. Shallow water waves have a speed that is influenced by the depth of the water. However, they are capable of traveling at speeds exceeding 750 km/hr (Figure 10.4.1)!

When tsunamis approach shore, they exhibit similar behavior to regular waves: as the water depth decreases, the speed of the waves decreases and the height of the waves starts to climb. Contrary to common misconceptions, tsunamis do not reach the beach in the form of enormous, breaking waves. Due to the lengthy wavelength, the height of these waves cannot surpass 1/7 of their wavelength, resulting in the absence of curling or breaking. Instead, they typically make landfall as abrupt inundations of water, resulting in a rapid and significant elevation in the sea level, akin to a massive tidal surge. The wave may require several minutes to traverse, during which the sea level might ascend up to 40 meters over its typical height.

Frequent tsunamis of significant magnitude occur every 2-3 years, while infrequent but highly destructive occurrences take place every 15-20 years. The most catastrophic tsunami in terms of casualties occurred as a result of a magnitude 9 earthquake in Indonesia in 2004 (Figure 10.4.2). This earthquake generated waves that reached heights of up to 33 meters and caused the death of over 230,000 individuals in Indonesia, Thailand, and Sri Lanka. In 2011, a seismic event of 9.0 magnitude occurred in Japan, causing a tsunami with maximum wave heights reaching 40.5 meters. This catastrophic event led to the unfortunate loss of over 18,000 lives. This seismic event also resulted in the Fukishima nuclear incident and shifted Japan approximately 8 inches closer to the United States.

Figure 10.4.2 A village in Sumatra following the Indonesia Tsunami in December 2004 (U.S. Navy photo by Photographer's Mate 2nd Class Philip A. McDaniel [Public domain via Wikimedia Commons]).

11 Chapter | TIDES

11.1 TIDAL FORCES

The foundation of our contemporary comprehension of tide formation originates from Isaac Newton's Law of Universal Gravitation, which asserts that there exists a gravitational attraction between any two things. The force's magnitude is directly proportional to the masses of the objects and inversely proportional to the square of the distance between them, as stated in the equation depicted in Figure 11.1.1.

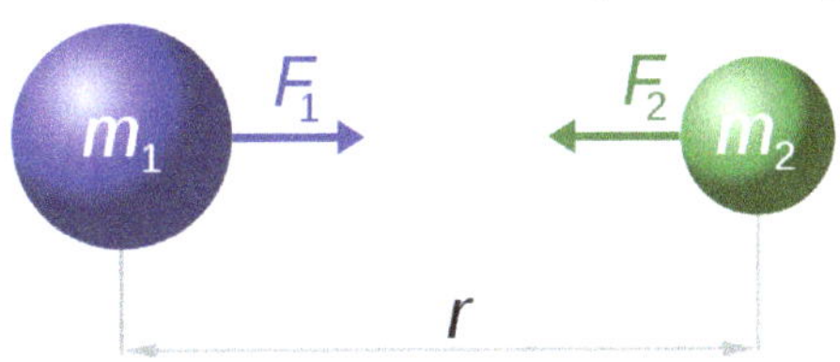

$$F_1 = F_2 = G\frac{m_1 \times m_2}{r^2}$$

Figure 11.1.1 Newton's Law of Universal Gravitation. The gravitational force between two objects (F) is calculated as the product of the two masses (m_1 and m_2) divided by the distance between them (r) squared. G is the universal gravitational constant; 6.67408×10^{-11} m^3 kg^{-1} s^{-2} (By I, Dennis Nilsson [CC BY 3.0], via Wikimedia Commons).

In the case of tides, there are a few other factors that modify this equation so that the distance (r) is cubed rather than squared, giving distance an even greater impact on tidal forces. But for our purposes, the important lesson is that the greater the masses of the objects, the greater the gravitational force, and the farther the objects are from each other, the weaker the force.

Such a gravitational force exists between the Earth and moon, attempting to pull them towards each other. Since the water covering Earth is fluid (unlike the solid land that is more resistant to tidal forces), this gravitational force pulls water towards the moon, creating a "bulge" of water on the side of the Earth facing the moon (Figure 11.1.2). This bulge always faces the moon, while the Earth rotates through it; the regions of Earth moving through the bulge experience a high tide, while those parts of the Earth away from the bulge experience a low tide.

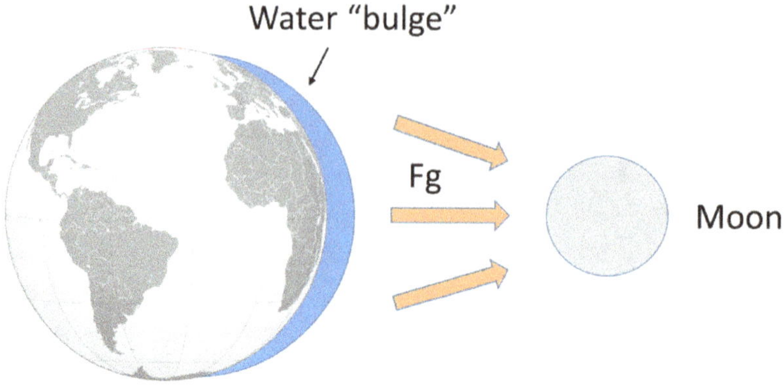

Figure 11.1.2 Gravitational forces between the Earth and moon cause a bulge of water to appear on the side of the Earth facing the moon (PW).

If the tides were this simple, everywhere on Earth would see one high tide per day, as there would only be a bulge of water on the side closest to the moon. However, if you have ever looked at tide charts, or lived near the ocean, you probably know that in most places there are two high tides and two low tides per day. Where is this second high tide "bulge" coming from?

The gravitational force between the Earth and moon might be expected to draw the two objects closer together, however, this is not happening. This is because the inward gravitational force is opposed by outward forces that keep the Earth and moon apart. The outward force is an **intertial force** created by the rotation of the Earth and moon. Contrary to popular belief, the moon is not simply rotating around the Earth; in fact, the Earth and moon are both rotating around each other. Imagine the Earth and moon as equal-sized objects revolving around a point at their center of mass. If both objects had the same mass, the center of rotation would be a point equidistant between the two objects. But since the mass of the Earth is 82 times greater than the mass of the moon, the center of revolution must be closer to the Earth. As an analogy, think about two people on a see-saw. If the people are of roughly equal size, they can sit on either end of the see-saw at it will rotate around a point at equal distance between them. But if the two people have very different masses, such as a large adult and a small child, the larger person must move closer to the pivot point for the see-saw to rotate effectively. In the same way, the center of rotation between the Earth and the moon (the **barycenter**) must be located closer to the Earth. In fact, the center of rotation lies *within* the Earth, about 1600 km below the surface. As the Earth and moon rotate around the barycenter, the moon travels much farther than the Earth, giving the impression that the moon is rotating around Earth (Figure 11.1.3).

The rotation of the Earth-moon system creates an outward inertial force, which balances the gravitational force to keep the two bodies in their orbits. The inertial force has the same magnitude everywhere on Earth, and is always directed away from the moon. Gravitational force, on the other hand, is always directed towards the moon, and is stronger on the side of the Earth closest to the moon. Figure 11.1.4 describes how these forces combine to create the tidal forces. At point O in the center of the Earth, the gravitational force (F_g) and the inertial force (F_r) are equal, and cancel each other out. On the side of Earth closest to the moon, the inward gravitational force (F_g) is greater than the outward inertial force (F_r); the net resulting force (A) is directed towards the moon, and creates a bulge of water on the side facing the moon. On the side of Earth opposite the moon, the outward inertial force is greater than the inward gravitational force; the net resulting force (C) is directed away from the moon, creating a water bulge directed away from the moon.

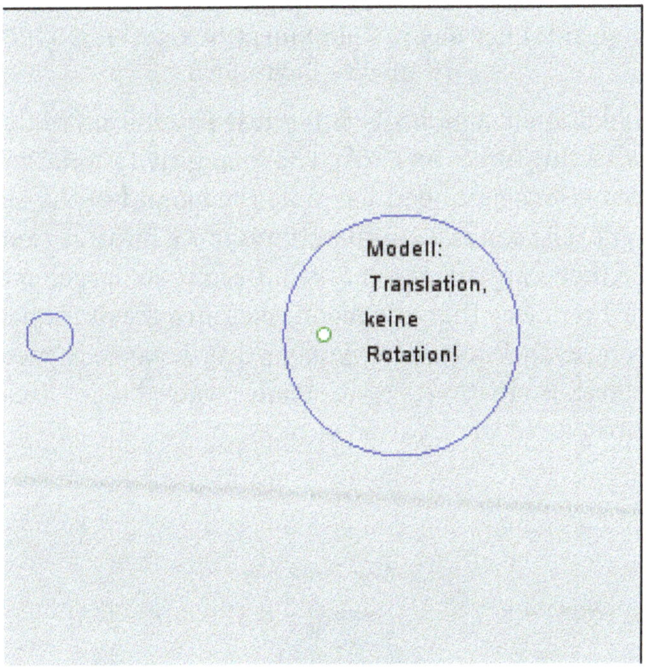

Figure 11.1.3 Rotation of the Earth and moon around the barycenter (white dot). Because of its larger mass, the center of rotation lies closer to the Earth than to the moon (By Modalanalytiker (Own work) [CC BY-SA 3.0 or GFDL (http://www.gnu.org/copyleft/fdl.html)], via Wikimedia Commons).

Now, as the Earth rotates through a 24 hour day, each region passes through two bulges, and experiences two high tides and two low tides per day. This represents Newton's **Equilibrium Theory of Tides,** where there are two high tides and two low tides per day, of similar heights, each six hours apart. But as with everything else in oceanography, reality is much more complex than this idealized situation.

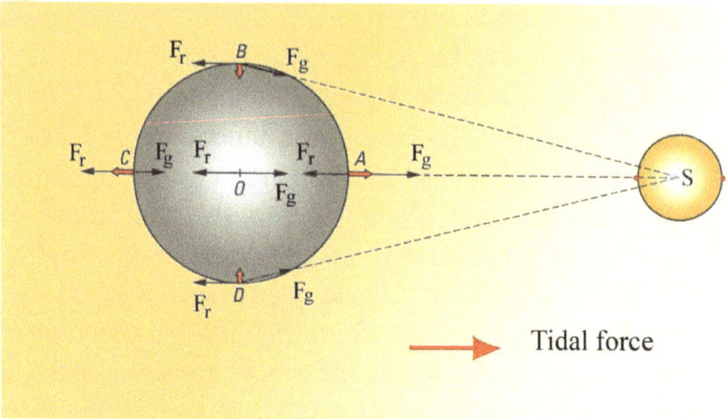

Figure 11.1.4 Gravitational force (F_g) are strongest closer to the moon and weaker opposite the moon. Inertial forces (F_r) are equal throughout the Earth and directed away from the moon. The tidal forces A and C are the result of the interaction between F_g and F_r and create water bulges on both sides of the Earth, leading to two high tides per day (Vitold Muratov (Own work) [CC BY-SA 3.0], via Wikimedia Commons).

Some of the additional complexity is because in addition to the moon, the sun also exerts tide-affecting forces on Earth. The solar gravitational and inertial forces arise for the same reasons described above for the moon, but the magnitudes of the forces are different. The sun is 27 million times more massive than the moon, but it is 387 times farther away from the Earth. Despite its larger mass, because the sun is so much farther away than the moon, the sun's gravitational forces are only about half as strong as the moon's (remember that distance is cubed in the gravity equation). The sun thus creates its own, smaller water bulges, independent of the moon's, that contribute to the creation of tides.

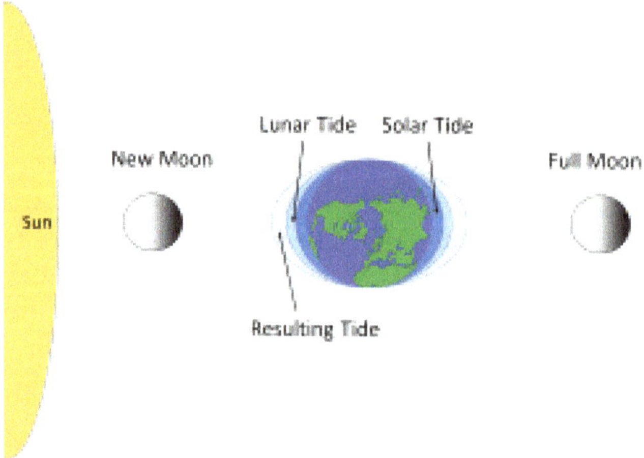

Figure 11.1.5 Spring tides with high tidal ranges occur when the solar and lunar tides are added together during full and new moons when the Earth, sun and moon are aligned (PW).

When the sun, Earth and moon are aligned, as occurs during new and full moons, the solar and lunar bulges are also aligned, and add to each other (constructive interference; see section 10.2) creating an especially high tidal range; high high tides and low low tides (Figure 11.1.5). This period of maximum tidal range is called a **spring tide**, and they occur every two weeks.

When the sun, Earth and moon are at 90° to each other, the solar and lunar bulges are out of phase, and cancel each other out (destructive interference). Now the tidal range is small, with low high tides and high low tides (Figure 11.1.6). These are **neap tides**, and occur every two weeks, when the moon is in its 1/4 and 3/4 phases (Figure 11.1.7).

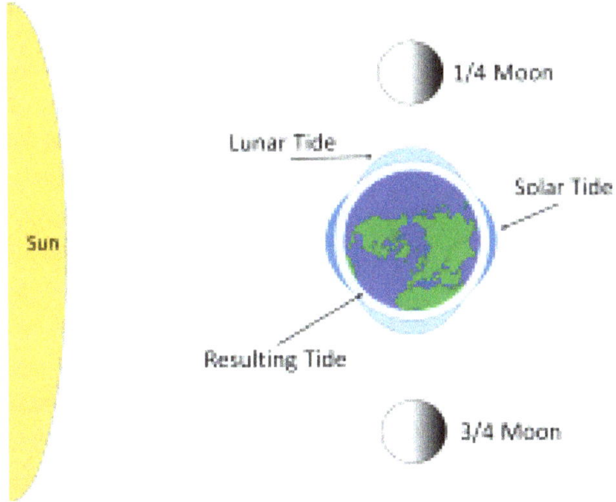

Figure 11.1.6 Neap tides are created during 1/4 and 3/4 moons when the Earth, sun and moon are perpendicular to each other. The solar and lunar tides cancel each other out, resulting in a small tidal range (PW).

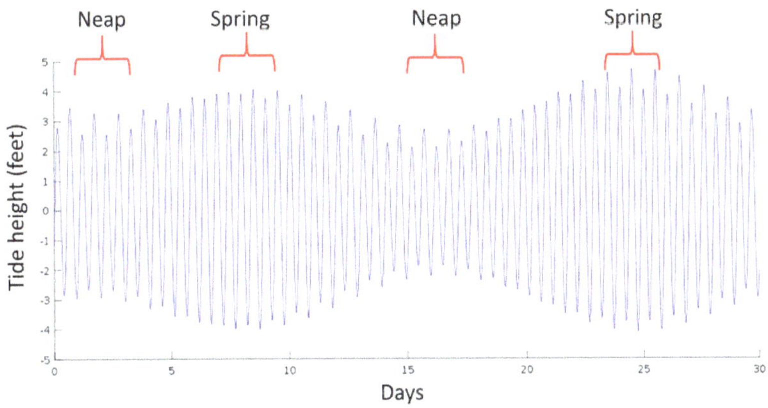

Figure 11.1.7 30 days of tidal data from Bridgeport, CT, USA, showing spring and neap tidal ranges (Modified by PW from Cody Logan (clpo13), Public Domain via Wikimedia Commons).

11.2 DYNAMIC THEORY OF TIDES

The Equilibrium Theory of tides predicts that each day there will be two high and two low tides, each one occurring at the same time day after day, with each pair producing tides of similar heights. While this view provides a basic explanation for the primary forces that generate the tides, it does not take into account such variables as the effects of the continents, the depth of the water, and many other factors. In all, there are almost 400 variables that must be incorporated into predicting the tides! The **Dynamic Theory** of tides takes these other factors into account, and shows that the tides are much more complicated and variable from place to place than the Equilibrium Theory would suggest. For example, some areas receive only one high and one low tide per day (see section 11.3). Furthermore, the tidal range varies greatly across the globe; in the Mediterranean Sea, there can be a difference of only 10 cm between high and low tides, while the Bay of Fundy in Canada experiences a tidal range of up to 17m (56 ft) every day (Figure 11.2.1).

Figure 11.2.1 Tidal range in the Bay of Fundy, Canada. Both photographs were taken on the same day in July 2003 (By Dylan Kereluk from White Rock, Canada (Flickr) [CC BY 2.0], via Wikimedia Commons).

Examination of any tide chart will show that the tides don't occur at same time each day; in fact, each tidal peak occurs about 50 minutes later than it did in the previous day. This is due to the orbit of the moon around the Earth. Imagine a high tide that occurs at a particular location (X) at 1:00 pm (Figure 11.2.2). The high tide occurs as location X moves through the bulge of water facing the moon. It will take the Earth 24 hours to complete one revolution, to bring location X back to site of the water bulge that caused that high tide. However, during those 24 hours, the moon has also moved as it orbits the Earth, so the high tide bulge has moved beyond its original location. The Earth thus has to rotate an additional distance for location X to reach the bulge and experience that same high tide. Because it takes the moon about 28 days to orbit the Earth, the moon gets "ahead" of the Earth's rotation by about 50 minutes per day. Therefore, it takes location X 24 hours and

50 minutes to rotate through the same tidal bulge, and as a result, the tidal peaks occur about 50 minutes later each day. In our example, an afternoon high tide at 1:00 pm on one day would be followed by a high tide at about 1:50 pm the following day. This 24 hour and 50 minute cycle is referred to as a **tidal day**.

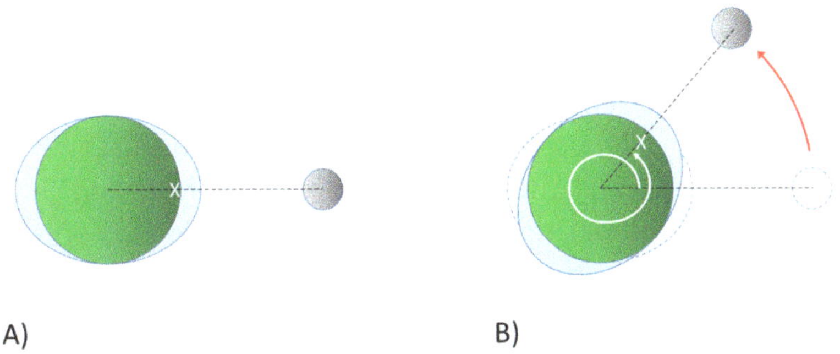

Figure 11.2.2 In A) a high tide is occurring at point X on Earth's surface. 24 hours later, X has made a complete rotation and is back in its original position. However, the moon has moved during that time (B), so X must travel an additional distance (white arrow) to once again become aligned with the moon and experience a high tide. For this reason, corresponding tides occur approximately 60 minutes later each day (PW).

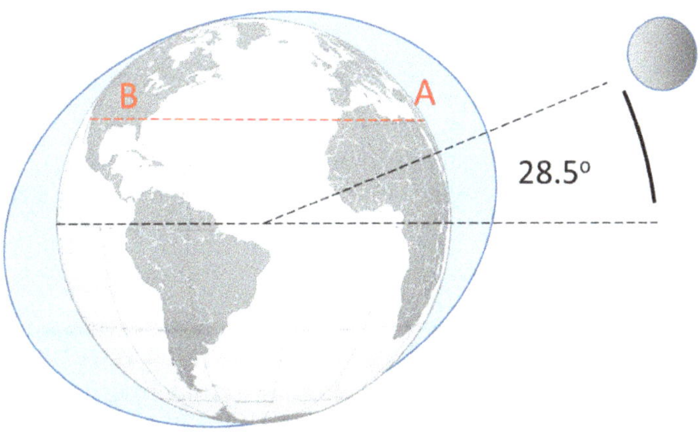

Figure 11.2.3 The effect of the moon declination on tide heights. The moon oscillates between 28.5° north and 28.5° south of the equator every two weeks, leading to uneven tidal heights each day at a particular latitude (PW).

The motion of the moon impacts the tidal cycles in other ways. As the moon orbits the Earth, its orbital plane is at an angle relative to the rotational plane of Earth. This angle, or **declination**, means than the moon fluctuates between an angle of 28.5° north of the equator, to 28.5° south of the equator roughly every two weeks (the cycle from maximum to minimum and back takes about 27 days). Figure 11.2.3 illustrates a case where the moon is at its maximum declination 28.5° north of the equator, creating its corresponding tidal maxima. A point on the Earth at the latitude indicated by the red line would experience two high tides as it rotated through 24 hours, at points A and B. But the two high tides would not be of equal heights; the high tide at A would be higher than the high tide at B. This helps create a mixed semi-diurnal tide; two high tides of different heights per day (see section 11.3).

Finally, the continents and the bottom topography of the oceans have an impact on the tides that are experienced in an area. Because the tides are essentially waves with extremely long wavelengths extending halfway across the Earth, they behave as shallow water waves, and they are influenced and refracted by the bottom contours, leading to regional tidal variations. When the tidal crests encounter land, they are are reflected, and the wave moves back out to sea, theoretically until it encounters another continent on the opposite side of the ocean basin. The crest is once again reflected, and the water oscillates back and forth as a standing wave across the ocean basin. However, because of the scale over which these tidal waves move, we must take into account the influence of the Coriolis Effect.

As the tidal crest is reflected back across the ocean basin, its path is deflected by the Coriolis force; to the right in the Northern Hemisphere, and to the left in the Southern Hemisphere (see section 9.1). Using the Northern Hemisphere as an example, imagine a tidal crest that has reached land on the western side of an ocean basin. It would have a tendency to be reflected and move across the basin towards the east. But the Coriolis force deflects the movement to the right, causing the crest to instead head south. When the crest hits land in the south, it would now tend to reflect towards the north, but once again the Coriolis deflection to the right kicks in, and the wave instead moves to the east. From the east the reflected wave is deflected to the north, and so on. The result of all of this is that instead of a simple standing wave moving back and forth across the ocean, the tidal crest follows a circular pattern around the ocean basin, counterclockwise in the Northern Hemisphere and clockwise in the Southern Hemisphere. This is analogous to shaking a pan full of water in a circular manner, and watching the water follow a similar circular path as it sloshes around inside. This large scale circular rotation pattern of tides is called **amphidromic circulation** (Figure 11.2.4). The rotation occurs around a central **amphidromic point** or **node**, that shows little tidal variation, while the largest tidal ranges occur on the edges of the circulation pattern.

In Figure 11.2.4 the amphidromic points are indicated by the dark blue areas where the white lines converge, like spokes from a bicycle wheel, and the dark red and brown areas show the regions of maximum tidal heights. The tidal maxima will rotate around the amphidromic points, taking about 12 hours for a complete rotation, leading to two high and two low tides per day in many places. If a tidal maximum is occurring along one of the white lines in Figure 11.2.4 at a certain time in the Northern Hemisphere, one hour later that high tide will have moved to the white line to the left (counterclockwise), and so on until it completes a rotation. In the Southern Hemisphere, the tide will move to the line to the right for clockwise rotation.

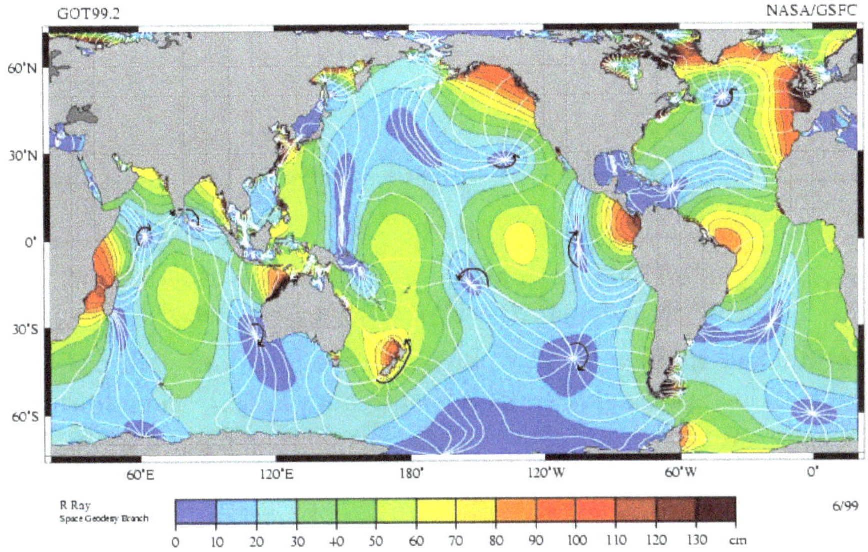

Figure 11.2.4 Amphidromic circulation. Amphidromic points are represented where the white lines converge in areas of minimal tidal range. Tidal crests rotate around the amphidromic points, counterclockwise in the Northern Hemisphere and clockwise in the Southern Hemisphere. See text for details (NASA, Public Domain via Wikimedia Commons).

The result of all of these variables is that the tides will not always occur twice each day, at the same time and with equal heights as the Equilibrium Theory of tides may suggest. Instead, each region of the oceans has a unique set of factors that contribute to the types of tides it will experience. The major types of tides are discussed in the next section.

11.3 TIDE CLASSIFICATION

With so many variables playing a role in the production of tides, it is understandable that not every place on Earth will experience exactly the same tidal conditions. There are three primary classifications for tides, depending on the number and relative heights of tidal cycles per day.

A **diurnal tide** consists of only one high tide and one low tide per day (Figure 11.3.1). "Diurnal" refers to a daily occurrence, so a situation where there is only one complete tidal cycle per day is considered a diurnal tide. Diurnal tides are common in the Gulf of Mexico, along the west coast of Alaska, and in parts of Southeast Asia.

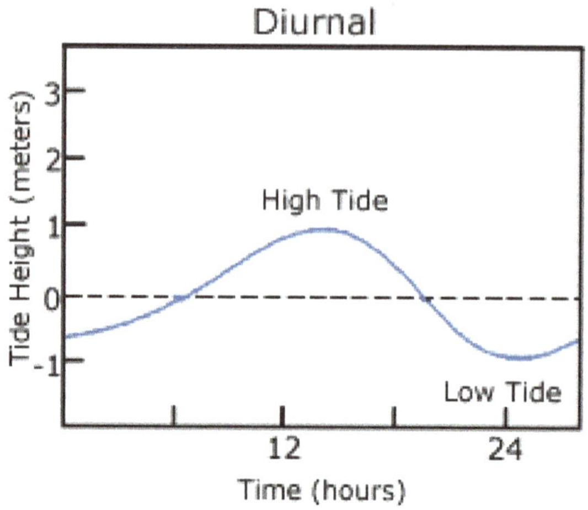

Figure 11.3.1 A diurnal tide, with one high and one low tide per day (By NOAA [Public domain], via Wikimedia Commons).

A **semidiurnal tide** exhibits two high and two low tides each day, with both highs and both lows of toughly equal height (Figure 11.3.2). "Semidiurnal" means "half of a day"; one tidal cycle takes half of a day, therefore there are two complete cycles per day. Semidiurnal tides are common along the east coasts of North America and Australia, the west coast of Africa, and most of Europe.

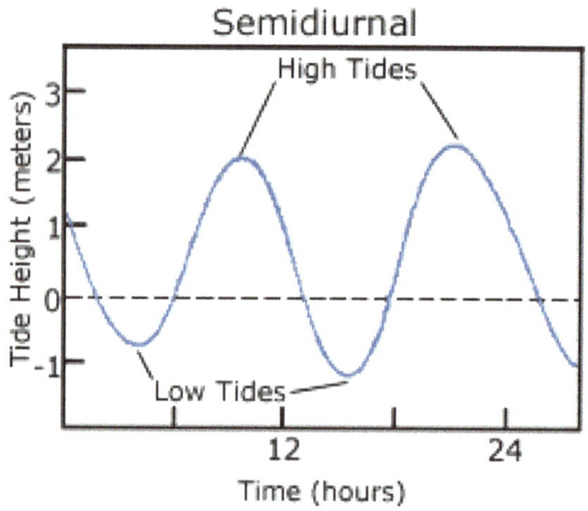

Figure 11.3.2 A semi-diurnal tide, with two high and two low tides per day, each of roughly equal heights (By NOAA [Public domain], via Wikimedia Commons).

Mixed semidiurnal tides (or **mixed tides**), have two high tides and two low tides per day, but the heights of each tide differs; the two high tides are of different heights, as are the two low tides (Figure 11.3.3). The differences in height may be the result of amphidromic circulation, the angle of the moon, or any of the other variables discussed in section 11.2. Mixed semidiurnal tides are found along the Pacific coast of North America.

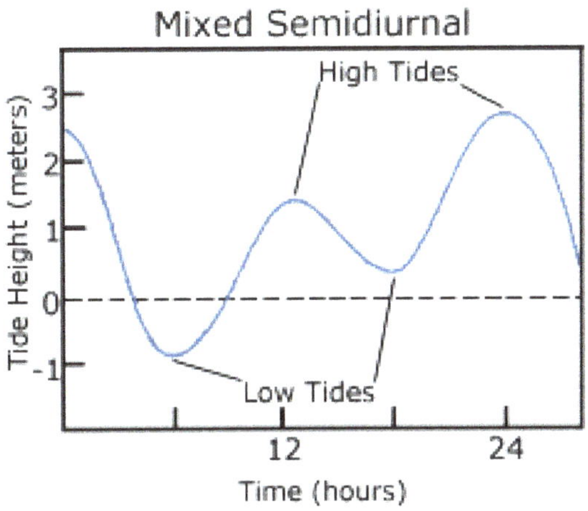

Figure 11.3.3 A mixed semi-diurnal tide, with two high and two low tides per day, each with a different height (By NOAA [Public domain], via Wikimedia Commons).

Figure 11.3.4 shows the distribution of the various tide types throughout the world.

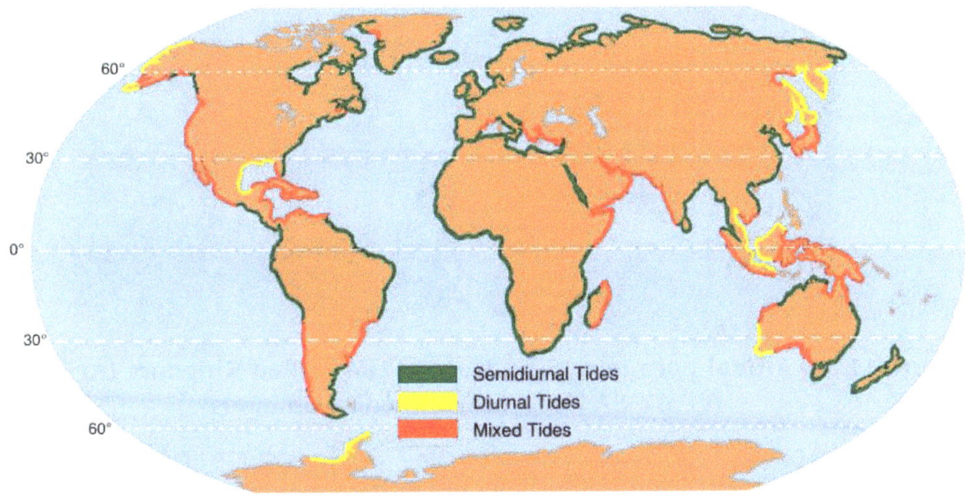

Figure 11.3.4 Global distribution of the different types of tides (By KVDP (Own work) [Public domain], via Wikimedia Commons).

Tidal Currents

Tidal currents result from the motion of water induced by the ebb and flow of the tide. As the water level rises, it penetrates a specific area, leading to the creation of a flood current. When the water level lowers and the movement of water retreats, it creates an ebb current. Slack water, or slack tides, refers to the period when there is no net water flow when the tide shifts from high to low.

The intensity of a tidal current is governed by the volume of water that enters and exits during each tidal cycle (known as the tidal volume or tidal prism), as well as the size of the area through which the water flows. A significant amount of water moving in a large area might create a weak current because the water is spread out over a wide space. In contrast, a limited area can produce a strong tidal current even with a small amount of water movement, because the water is focused and propelled into a confined space. The most potent tidal currents will arise when a substantial disparity in tide elevation traverses a restricted area.

Tidal bores occur at the point where rivers and the ocean meet. A tidal bore occurs when the strength of the incoming tidal current is greater than that of the river outflow. It appears as a wave or a moving barrier of water that moves upstream with the oncoming tide (Figure 11.3.5).

Figure 11.3.5 A tidal bore near Silverdale in the United Kingdom (Arnold Price [CC BY-SA 2.0], via Wikimedia Commons).

Chapter 12 OCEAN SEDIMENTS

12.1 CLASSIFICATION OF SEDIMENTS

Sediment is the collective word for minuscule rock particles and other substances that descend to the ocean floor, gradually settling and accumulating at the bottom. Every part of the bottom has sediment, albeit with diverse origins and compositions. The quantity of deposited sediment might significantly range across different locations. On a global scale, the thickness of ocean sediments typically measures around 1 kilometer. However, in regions with significant sediment buildup, the thickness can surpass 15 kilometers (Figure 12.1.1).

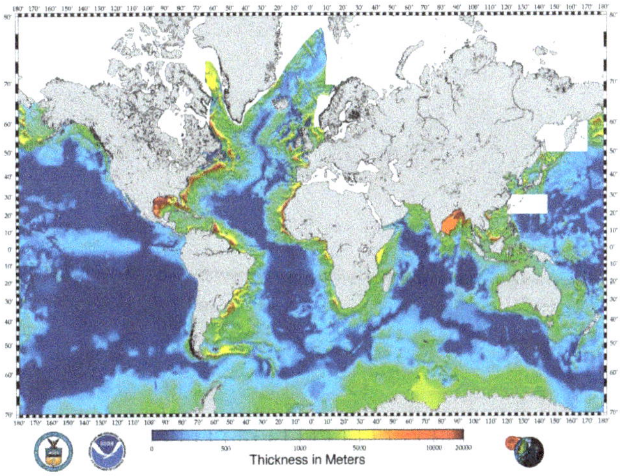

Figure 12.1.1 Total sediment thickness of the world ocean (By Divins, D.L., NGDC Total Sediment Thickness of the World's Oceans & Marginal Seas, http://www.ngdc.noaa.gov/mgg/sedthick/sedthick.html (http://www.ngdc.noaa.gov/mgg/image/sedthick9.jpg) [Public domain], via Wikimedia Commons).

These areas encompass locations in close proximity to river mouths with significant sediment discharge, as well as passive margins adjacent to continents where sediment has accumulated over millions of years. In contrast, sediments are scarce around divergent plate boundaries where new oceanic crust is being generated, as the crust is too young to accumulate significantly (see to section 4.5). Additionally, sediments are also limited in the central oceans that are distant from any substantial sources of sediment.

Over the course of millions of years, the sediments have the potential to undergo lithification, transforming into sedimentary rock. More than 50% of the rock found on the continents is sedimentary rock that was initially formed in ancient oceans and later raised up by plate tectonics. Several towering mountains, such as Mt. Everest, are comprised of rock formations that house fossils of marine organisms. The rocks originated as marine sediments, which then underwent lithification and were uplifted during the process of orogenesis. Ocean sediments can be classified in various ways, with some of the most prevalent criteria being the texture, composition, and provenance of the sediment.

Texture

Sediment texture can be examined through several variables. The first is **grain size.** Sediments are classified by particle size, ranging from the finest clays (diameter <0.004 mm) to the largest boulders (> 256 mm)(Figure 12.1.2). Among other things, grain size represents the conditions under which the sediment was deposited. High energy conditions, such as strong currents or waves, usually results in the deposition of only the larger particles as the finer ones will be carried away. Lower energy conditions will allow the smaller particles to settle out and form finer sediments.

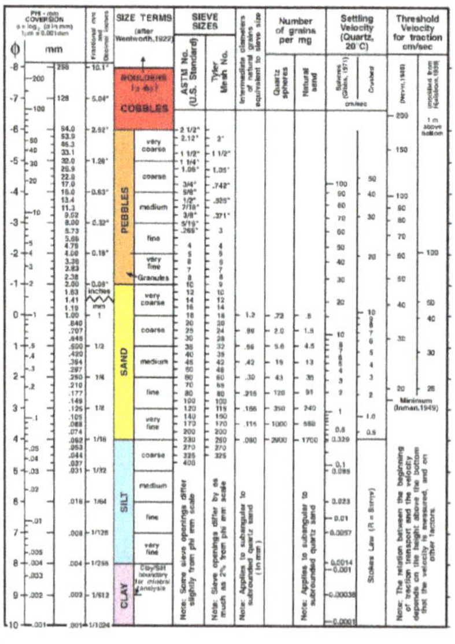

Figure 12.1.2 Wentworth grain size chart for classifying sediments (By Jeffress Williams, Matthew A. Arsenault, Brian J. Buczkowski, Jane A. Reid, James G. Flocks, Mark A. Kulp, Shea Penland, and Chris J. Jenkins, USGS [Public domain], via Wikimedia Commons).

Sorting is another way to categorize sediments. Sorting refers to how uniform the particles are in terms of size (Figure 12.1.3). If all of the particles are of a similar size, such as in beach sand, the sediment is well-sorted. If the particles are of very different sizes, the sediment is poorly sorted, such as in glacial deposits.

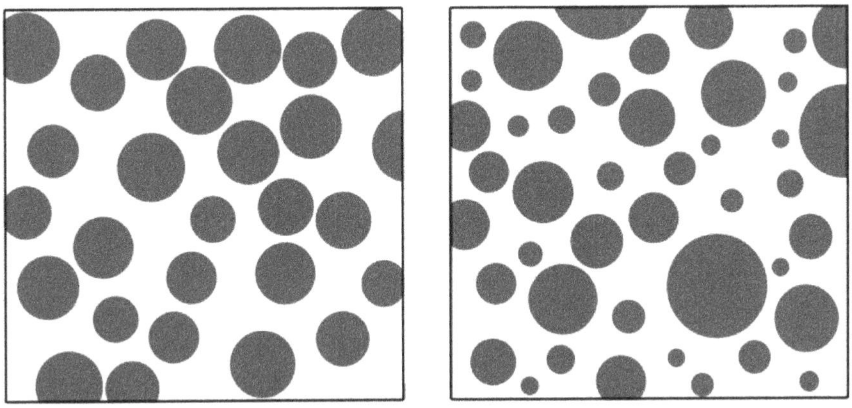

Figure 12.1.3 Well-sorted sediments (left) have particles that are all of a similar size. Poorly sorted sediments (right) consist of particles of a wide range of sizes (Woudloper (Own work) [CC BY-SA 3.0], via Wikimedia Commons).

Ultimately, the composition of sediment can be characterized by its maturity, which refers to the duration that the particles have been carried by water or other agents. The level of maturity can be determined by the extent of particle rounding, the degree of sorting, and the sediment composition. When it comes to rounding, the level of maturity of sediment directly affects the roundness of its particles due to the gradual abrasion they undergo throughout time. A high level of sorting is indicative of maturity, as it suggests that smaller particles have been gradually removed over time, and that a consistent quantity of energy has transported particles of similar sizes over the same distance. In sediments formed from rock particles, the quartz content increases as the sediment ages and becomes more mature. Quartz is a prevalent mineral found in rocks on Earth, known for its exceptional hardness and resistance to abrasion. Over time, particles composed of different substances erode, leaving behind only quartz. Beach sand is an extremely developed sediment, consisting mainly of quartz, with rounded particles that are well-sorted and of comparable size.

Sources of sediments

Sediments are also classified based on their source of origin. There are four main categories for the origin of marine sediments:

- **Lithogenous** sediments are derived from preexisting rock. They are also called **terrigenous** sediments since most of it comes from the land masses and makes its way into the ocean.

- **Biogenous** sediments are composed of the remains of marine organisms.
- **Hydrogenous** sediments are formed when materials that are dissolved in water precipitate out and form solid particles.
- **Cosmogenous** sediments are derived from extraterrestrial sources.

The next few sections will address each of these sediment types in more detail.

12.2 LITHOGENOUS SEDIMENTS

Lithogenous or terrigenous sediment consists predominantly of minute fragments of previous rocks that have been transported into the ocean. The sediments found on the ocean floor can encompass a wide variety of particle sizes, ranging from minuscule clays to enormous boulders, and are present in nearly all areas of the ocean floor. Lithogenous sediments are formed on land by the process of weathering, which involves the breakdown of rocks and minerals into smaller particles due to the effects of wind, rain, water movement, temperature changes, ice-induced cracking, and other erosive processes. Subsequently, these diminutive weathered fragments are conveyed to the oceans via a multitude of mechanisms:

- **Streams and rivers:** Various forms of runoff deposit large amounts of sediment into the oceans, mostly in the form of finer-grained particles (Figure 12.2.1). About 90% of the lithogenous sediment in the oceans is though to have come from river discharge, particularly from Asia. Most of this sediment, especially the larger particles, will be deposited and remain fairly close to the coastline, however, smaller clay particles may remain suspended in the water column for long periods of time and may be transported great distances from the source.

Figure 12.2.1 River discharge in the Yukon Delta, Alaska. The pale color demonstrates the large amounts of sediment released into the ocean via the rivers (By Jesse Allen and Robert Simmon (NASA Earth Observatory) [Public domain], via Wikimedia Commons).

- **Wind**: Wind-borne (**aeolian**) transport can take small particles of sand and dust and move them thousands of kilometers from the source. These small particles can fall into the ocean when the wind dies down, or can serve as the nuclei around which raindrops or snowflakes form. Aeolian transport is particularly important near desert areas (Figure 12.2.2).

Figure 12.2.2 A plume of wind-borne particles from Sudan (left) blow over the Red Sea (By NASA (http://visibleearth.nasa.gov/view_rec.php?id=5645) [Public domain], via Wikimedia Commons).

- **Glaciers and ice rafting**: As glaciers grind their way over land, they pick up lots of soil and rock particles, including very large boulders, that get carried by the ice. When the glacier meets the ocean and begins to break apart or melt, these particles get deposited (Figure 12.2.3). Most of the deposition will happen close to where the glacier meets the water, but a small amount of material is also transported longer distances by rafting, where larger pieces of ice drift far from the glacier before releasing their sediment.

Figure 12.2.3 When glaciers reach the sea they can break apart, depositing their sediments into the ocean, including very large pieces of rock (By Ianaré Sévi (Own work) [CC BY-SA 3.0], via Wikimedia Commons).

- **Gravity:** Landslides, mudslides, avalanches, and other gravity-driven events can deposit large amounts of material into the ocean when they happen close to shore.
- **Waves:** Wave action along a coastline will erode rocks and will pull loose particles from beaches and shorelines into the water.
- **Volcanoes:** Volcanic eruptions emit vast amounts of ash and other debris into the atmosphere, where it can then be transported by wind to eventually get deposited in the oceans (Figure 12.2.4).

Figure 12.2.4 Eruption of the Mayon Volcano, Philippines, in 1984. Much of the material spewed from a volcanic eruption may eventually make its way into the oceans (By C.G. Newhall [Public domain], via Wikimedia Commons).

- **Gastroliths:** An interesting, although relatively minor avenue for the transport of lithogenous sediments to the ocean is in the form of gastroliths. Gastrolith means "stomach stones" and comes from the fact that many animals, including seabirds, pinnipeds, and some crocodiles will deliberately swallow stones in one area and regurgitate them in another. Often these stones swallowed on land will be regurgitated at sea. Why swallow stones? Possible explanations include using the stones to help grind up food in the stomach, to act as ballast to aid in buoyancy regulation, or to fill the stomach and reduce feelings of hunger during fasting periods on shore.

The majority of these mechanisms deposit lithogenous sediment in close proximity to the shoreline. The sediment particles can subsequently be carried to distant locations by the action of waves and currents, perhaps reaching the deep ocean floor after escaping the continental shelf.

Composition

Lithogenous sediments typically mirror the composition of the materials they originated from, therefore being predominantly composed of the primary minerals found in terrestrial rock. The components encompassed in this are quartz, feldspar, clay minerals, iron oxides, and terrestrial organic stuff. Quartz, which is mostly made up of silicon dioxide and is a key element of glass, is a prevalent mineral found in almost all types of rocks. It possesses a high level of durability against abrasion, making it a prominent constituent of lithogenous deposits, such as sand.

12.3 BIOGENOUS SEDIMENTS

Biogenous sediments come from the remains of living organisms that settle out as sediment when the organisms die. It is the "hard parts" of the organisms that contribute to the sediments; things like shells, teeth or skeletal elements, as these parts are usually mineralized and are more resistant to decomposition than the fleshy "soft parts" that rapidly deteriorate after death.

Macroscopic sediments contain large remains, such as skeletons, teeth, or shells of larger organisms. This type of sediment is fairly rare over most of the ocean, as large organisms don't die in enough of a concentrated abundance to allow these remains to accumulate. One exception is around coral reefs; here there is a great abundance of organisms that leave behind their remains, in particular the fragments of the stony skeletons of corals that make up a large percentage of tropical sand.

Microscopic sediment consists of the hard parts of microscopic organisms, particularly their shells, or **tests.** Although very small, these organisms are highly abundant and as they die by the billions every day their tests sink to the bottom to create biogenous sediments. Sediments composed of microscopic tests are far more abundant than sediments from macroscopic particles, and because of their small size they create fine-grained, mushy sediment layers. If the sediment layer consists of at least 30% microscopic biogenous material, it is classified as a biogenous **ooze.** The remainder of the sediment is often made up of clay.

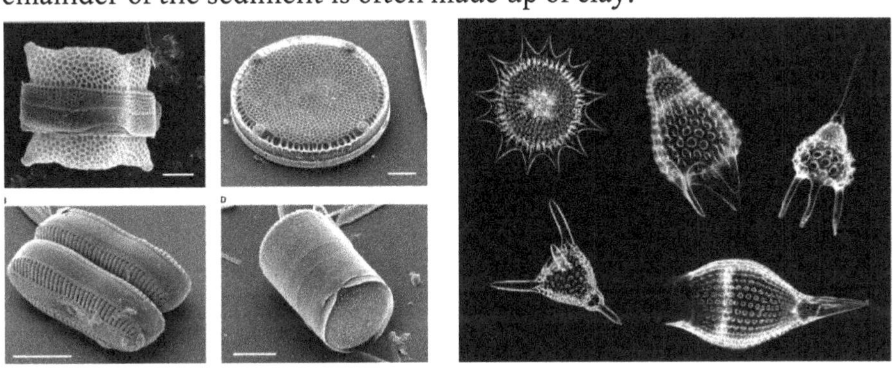

Figure 12.3.1 Various diatom (left) and radiolarian (right) tests (Diatom images courtesy of Mary Ann Tiffany, San Diego State University [CC BY 2.5], via Wikimedia Commons; radiolarian images by Andreas Drews, https://pxhere.com/en/photo/239774, [CC by 2.0]).

The primary sources of microscopic biogenous sediments are unicellular algaes and protozoans (single-celled amoeba-like creatures) that secrete tests of either calcium carbonate ($CaCO_3$) or silica ($SiO_{2)}$. Silica tests come from two main groups, the **diatoms** (algae) and the **radiolarians** (protozoans) (Figure 12.3.1).

Diatoms are important members of the phytoplankton, the small, drifting algal photosynthesizers. A diatom consists of a single algal cell surrounded by an elaborate silica shell that it secretes for itself. Diatoms come in a range of shapes, from elongated, pennate forms, to round, or centric shapes that often have two halves, like a Petri dish (Figure 12.3.1 *left*). In areas where diatoms are abundant, the underlying sediment is rich in silica diatom tests, and is called **diatomaceous earth** (see box below).

What use are diatoms?

Diatoms are a vital piece of the global ecosystem for their role in oceanic primary production and the creation of much of the oxygen that organisms breathe. But diatoms are also important for many industrial and agricultural applications. Because of the very fine grain size, and the lattice-like structure of the diatom tests, diatomaceous earth has been used as a filtering agent in things like swimming pool filters and beer brewing. The microscopic tests have been added as an abrasive to toothpaste, facial cleansers and household cleaning agents. Alfred Nobel used diatomaceous earth to stabilize nitroglycerine in the production of dynamite. Diatomaceous earth also displays insecticide properties by stimulating dehydration in insects. It is marketed for this purpose in agriculture, as well as for household use to combat ants, cockroaches, and bedbugs. "Food grade" diatomaceous earth has also entered the market, with proponents touting a range of health benefits arising from its consumption. That's a pretty impressive range of uses from a microscopic algae!

Radiolarians are planktonic protozoans (making them part of the zooplankton), that like diatoms, secrete a silica test. The test surrounds the cell and can include an array of small openings through which the radiolarian can extend an amoeba-like "arm" or pseudopod (Figure 12.3.1 *right*). Radiolarian tests often display a number of rays protruding from their shells which aid in buoyancy. Oozes that are dominated by diatom or radiolarian tests are called **siliceous oozes.**

Like the siliceous sediments, the calcium carbonate, or calcareous sediments are also produced from the tests of microscopic algae and protozoans; in this case the **coccolithophores** and **foraminiferans.** Coccolithophores are single-celled planktonic algae about 100 times smaller than diatoms. Their tests are composed of a number of interlocking $CaCO_3$ plates (coccoliths) that form a sphere surrounding the cell (Figure 12.3.2 *left*). When coccolithophores die the individual plates sink out and form an ooze. Over time, the coccolithophore ooze lithifies to becomes chalk.

The famous White Cliffs of Dover in England are composed of coccolithophore-rich ooze that turned into chalk deposits (Figure 12.3.2 *right*).

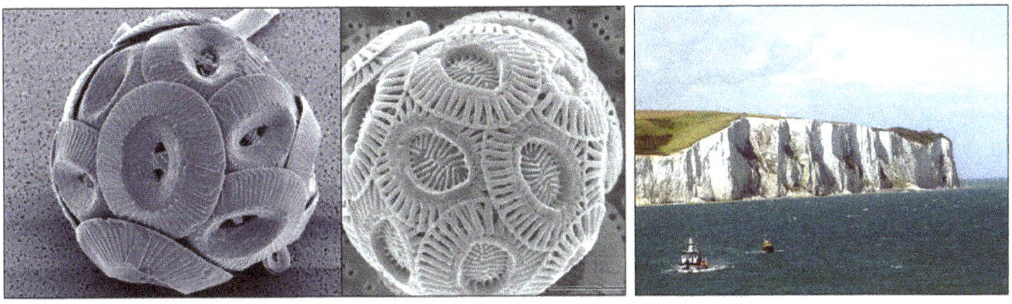

Figure 12.3.2 (Left) coccolithophore tests (left; By Richard Lampitt, Jeremy Young, The Natural History Museum, London (http://planktonnet.awi.de/); center; by Alison R. Taylor (University of North Carolina Wilmington Microscopy Facility) (PLoS Biology, June 2011, Cover ([1])) [Both images CC BY 2.5], via Wikimedia Commons). (Right); the White Cliffs of Dover (Immanuel Giel (Own work) [CC BY-SA 3.0], via Wikimedia Commons).

Foraminiferans (also referred to as "forams") are protozoans whose tests are often chambered, similar to the shells of snails. As the organism grows, is secretes new, larger chambers in which to reside. Most foraminiferans are benthic, living on or in the sediment, but there are some planktonic species living higher in the water column. When coccolithophores and foraminiferans die, they form **calcareous oozes.**

Figure 12.3.3 Foraminifera tests collected from a beach in Myanmar (By Psammophile [GFDL (http://www.gnu.org/copyleft/fdl.html) or CC BY-SA 3.0], via Wikimedia Commons).

Older calcareous sediment layers contain the remains of another type of organism, the **discoasters**; single-celled algae related to the coccolithophores that also produced calcium carbonate tests. Discoaster tests were star-shaped, and reached sizes of 5-40 μm across (Figure 12.3.4). Discoasters went extinct approximately 2 million years ago, but their tests remain in deep, tropical sediments that predate their extinction.

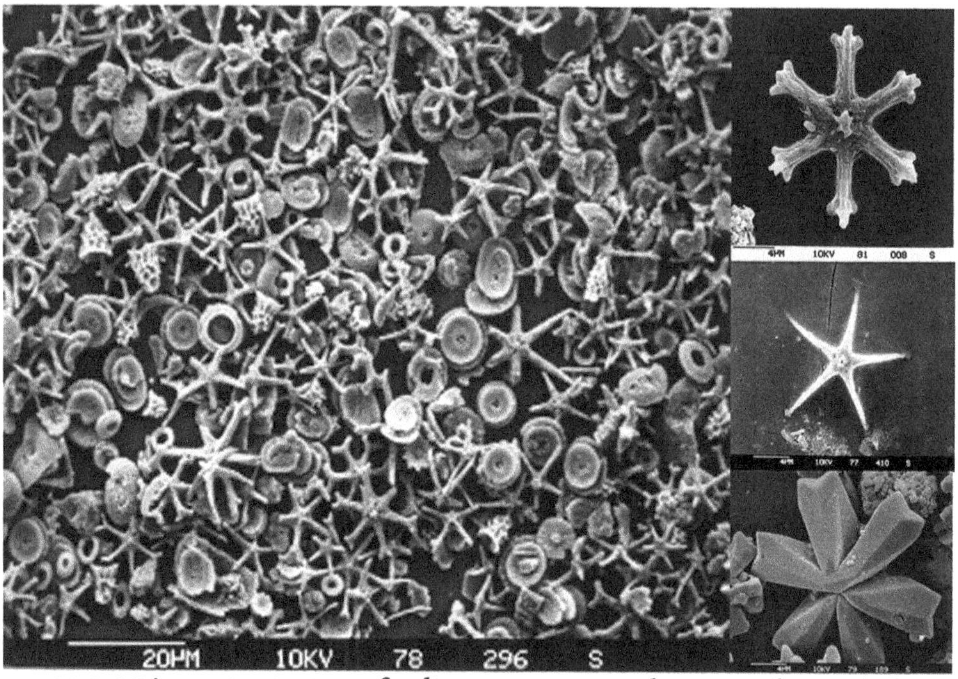

Figure 12.3.4 Discoaster tests. Left: discoster tests with assorted coccoliths. Top right *Discoaster surculus;* center right: *Discoaster pentaradiatus;* bottom right: *Discoaster surculus* (All images by Hannes Grobe (Own work) [CC BY 3.0], via Wikimedia Commons).

Because of their small size, these tests sink very slowly; a single microscopic test may take about 10-50 years to sink to the bottom! Given that slow descent, a current of only 1 cm/sec could carry the test as much as 15,000 km away from its point of origin before it reaches the bottom. Yet despite this, we find that the sediments in a particular location are well-matched to the types of organisms and degree of productivity that occurs in the water overhead. This means that the sediment particles must be sinking to the bottom at a much faster rate, so that they accumulate below their point of origin before the currents can disperse them. What is the mechanism for this increased sinking rate? Apparently most of the tests do not sink as individual particles; about 99% of them are first consumed by some other organism, and are then aggregated and expelled as large fecal pellets, which sink much more quickly and reach the ocean floor in only 10-15 days. This does not give the particles as much time to disperse, and the sediment below will reflect the production occurring near the surface. The increased rate of sinking through this mechanism is called the "fecal express."

Examining sediment analysis to reconstruct historical climatic patterns

As outlined in the opening to this chapter, examining marine sediments allows us to learn much about oceanographic and atmospheric processes, both past and present. Biogenous sediments are no exception, and they can allow us to reconstruct past climate history from oxygen isotope ratios.

Oxygen atoms exist in three forms, or isotopes, in ocean water: O^{16}, O^{17} and O^{18} (the number refers to the atomic masses of the isotopes). O^{16} is the most common form, followed by O^{18} (O^{17} is rare). O^{16} is lighter than O^{18}, so it evaporates more easily, leading to water vapor that has a higher proportion of O^{16}. During periods of cooler climate, water vapor condenses into rain and snow, which forms glacial ice that has a high proportion of O^{16}. The remaining seawater therefore has a relatively higher proportion of O^{18}. Marine organisms who incorporate dissolved oxygen into their shells as calcium carbonate will therefore have shells with a higher proportion of O^{18} isotope. In other words, the ratio of O^{16}:O^{18} in shells will be low during periods of colder climate.

When the climate warms, glacial ice melts, releasing O^{16} from the ice and returning it to the oceans, increasing the O^{16}:O^{18} ratio in the water. Now, when organisms incorporate oxygen into their shells, the shells will contain a higher O^{16}:O^{18} ratio. Scientists can therefore examine biogenous sediments, calculate the O^{16}:O^{18} ratios for samples of known ages, and from those ratios, infer the climate conditions under which those shells were formed. The same types of measurements can also be taken from ice cores; a decrease of 1 ppm O^{18} between ice samples represents a decrease in temperature of 1.5° C.

12.4 HYDROGENOUS SEDIMENTS

*Methane hydrate section modified from "Physical Geology" by Steven Earle**

As observed in section 5.3, seawater comprises a multitude of distinct dissolved compounds. Periodically, chemical reactions occur that result in the formation of solid particles, which subsequently gather as hydrogenous silt. Typically, these reactions are initiated by alterations in circumstances, such as variations in temperature, pressure, or pH, leading to a decrease in the solubility of a chemical. Hydrogenous sediment is rather scarce in the ocean when compared to lithogenous or biogenous sediments. However, there are still some intriguing variations present.

The topic of hydrothermal vents was addressed in section 4.11. Remember that in these systems, seawater infiltrates the seafloor and undergoes intense heating from magma before being released out the vent. The superheated water is rich in dissolved compounds. When it comes into contact with the cold seawater after leaving the vent, these substances solidify, primarily as metal sulfides. These particles constitute the "smoke" that emerges from a vent and may eventually deposit as hydrogenous silt on the seafloor (Figure 12.4.1).

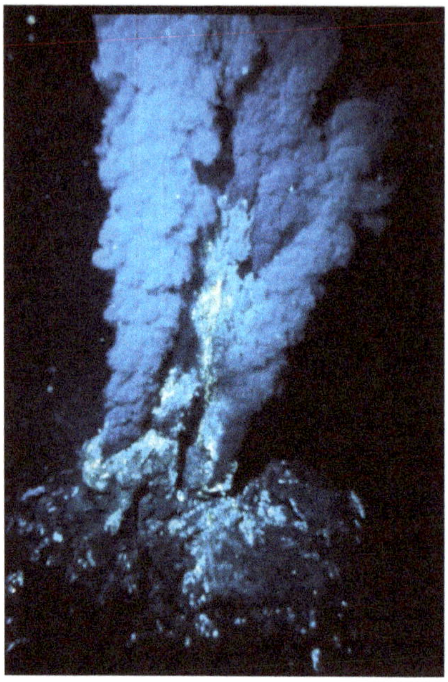

Figure 12.4.1 A "black smoker" hydrothermal vent. The "smoke" consists of dissolved particles that precipitate into solids when exposed to colder water (NOAA, http://www.photolib.noaa.gov/htmls/nur04506.htm).

Manganese nodules are spherical masses composed of manganese and several other metals that develop on the ocean floor. They typically have a diameter of 3-10 cm, however occasionally they can grow as large as 30 cm (Figure 12.4.2). The nodules develop through a process like the formation of pearls, where successive layers are gradually added around a core object, resulting in the gradual growth of the nodule. The composition of the nodules may exhibit considerable variation depending on their location and the circumstances of their development, but they are often characterized by a prevalence of manganese- and iron oxides. In addition, they may also include lesser quantities of other metals such as copper, nickel, and cobalt. Manganese nodules undergo an extremely slow geological process, growing at a rate of only a few millimeters per million years. As a result, these nodules only develop in regions with minimal lithogenous or biogenous sediment buildup, as any additional sediment deposition would rapidly bury the nodules and hinder their growth. As a result, manganese nodules are typically found in remote regions of the middle ocean, far away from important sources of lithogenous or biogenous materials. In these places, they can occasionally amass in great quantities on the seafloor (see Figure 12.4.2 on the right). Due to the presence of rich metals, there has been considerable interest in mining the nodules for many years. However, most of the attempts have only reached the exploratory phase. Several impediments have hindered the extensive retrieval of nodules, such as the exorbitant expenses associated with deep sea mining endeavors, political disputes regarding mining privileges, and environmental apprehensions related to the extraction of these finite resources.

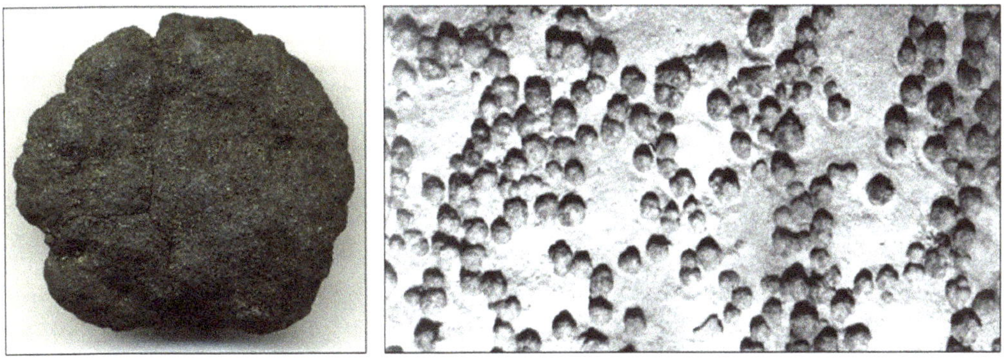

Figure 12.4.2 (Left) manganese nodule from the subtropical eastern Pacific Ocean. The nodule is 4.1 cm in diameter (By James St. John, https://www.flickr.com/photos/jsjgeology/15139986302/in/photostream/ [CC-BY 2.0]). (Right) field of manganese nodules on the seafloor (By United States Geological Survey [Public domain], via Wikimedia Commons).

Figure 12.4.3 Salt farmers harvesting salt left behind from the evaporation of seawater, Pak Thale, Ban Laem, Phetchaburi, Thailand (By JJ Harrison (Own work) [CC BY-SA 3.0], via Wikimedia Commons).

Evaporites are sediments of hydrogenous origin that result from the evaporation of seawater, causing the dissolved substances to solidify, primarily as halite (salt, NaCl). Seawater evaporation is the most ancient method of salt manufacturing for human consumption, and it continues to be practiced in the present day. Significant accumulations of halite evaporites are present in several locations, notably beneath the Mediterranean Sea. Approximately 6 million years ago, tectonic activities caused the Mediterranean Sea to become isolated from the Atlantic Ocean. As a result, the high temperatures caused a significant amount of water to evaporate, leading to the near-complete drying out of the Mediterranean. This event, referred to as the Messinian Salinity Crisis, resulted in the formation of extensive salt deposits. Subsequently, approximately 5.3 million years ago, the Mediterranean Sea saw a resurgence of water, resulting in the submergence of the halite deposits beneath further layers of sediment. Nevertheless, these deposits persist beneath the seafloor.

Oolites are minute, spherical particles that develop from the sequential deposition of substances surrounding a suspended particle. Typically, they consist of calcium carbonate, however they can also form from phosphates and other substances. Oolitic sand is formed through the accumulation of oolites, which are most commonly found in large quantities in the Bahamas (Figure 12.4.4).

Figure 12.4.4 Oolites from a beach on Joulter's Cay, The Bahamas (By Wilson44691 (Own work) [Public domain], via Wikimedia Commons).

Methane hydrates are a form of hydrogenous deposit that may have industrial use. Terrestrial erosion products contain a tiny amount of organic materials, primarily sourced from plants on land. Minute particles of this substance, along with additional organic substances derived from marine flora and fauna, amass in terrigenous sediments, particularly in close proximity to the shoreline within a few hundred kilometers. As sediment accumulates, the lower layers gradually heat up due to geothermal energy, prompting microorganisms to decompose the organic material there. Due to the lack of oxygen, known as anaerobic conditions, the metabolic process produces methane gas (CH4) as a by-product. The bacteria gradually release methane, which rises in the form of bubbles through the sediment towards the seafloor. Methane hydrate is formed when water and methane mix at depths ranging from 500 m to 1,000 m, where the bottom temperatures are close to 4°C. At depths ranging from a few meters to hundreds of meters below the seafloor, the temperature is sufficiently low for methane hydrate to remain stable. As a result, hydrates build within the sediment, as shown in Figure 12.4.5 on the left. Methane hydrate exhibits flammability due to the liberation of methane gas when subjected to heat (Figure 12.4.5, right). The methane contained in bottom sediments is a vast storehouse of fossil fuel energy. Despite the eagerness of energy businesses and governments to create methods for producing and selling methane, it is evident to those who comprehend the climate-change consequences of its extraction and use that pursuing such endeavors would be unwise.

Figure 12.4.5 (Left): Methane hydrate within muddy sea-floor sediment from an area offshore from Oregon (By Wusel007 (Own work) [GFDL (http://www.gnu.org/copyleft/fdl.html) or CC BY-SA 3.0], via Wikimedia Commons). (Right): Methane hydrate on fire (USGS, http://www.usgs.gov/blogs/features/files/2012/01/New-Image.jpg).

12.5 COSMOGENOUS SEDIMENTS

Cosmogenous silt originates from celestial sources and exists in two main forms: minuscule spherules and bigger fragments of meteors. Spherules consist primarily of silica or iron and nickel, and are believed to be expelled as meteors disintegrate upon entering the atmosphere. Meteor debris is the result of meteorites colliding with Earth. These intense collisions release particles into the atmosphere, which eventually return to Earth and contribute to the sediments. Meteor debris consists primarily of silica or iron and nickel, resembling spherules in shape. Tektites, microscopic glass droplets, are a fascinating type of debris resulting from these impacts. It is probable that they consist of silica from the Earth's surface that was expelled and melted during a meteorite collision, and subsequently hardened when it cooled upon reentry.

Figure 12.5.1 Tektite-like glass found near the Volkov River in western Russia (By James St. John [CC BY 2.0], via Wikimedia Commons).

Cosmogenous silt is quite uncommon in the ocean and typically does not amass in substantial quantities. Nevertheless, Earth is perpetually accumulating space dust from an ongoing shower of particles. Approximately 90% of cosmogenous debris disintegrates upon entering the Earth's atmosphere, whereas it is believed that between 5 and 300 tons of space dust settle on the Earth's surface daily.

12.6 SEDIMENT DISTRIBUTION

Having acquired knowledge on the many types of sediments present in the ocean, we can now shift our focus towards the mechanisms responsible for the prevalence of specific sediment types in particular areas. The rate of sediment deposition is contingent upon factors such as the quantity of material originating from the source, the proximity to the source, the duration during which sediment has been accumulating, the degree of sediment preservation, and the quantities of different sediment types being introduced into the system.

The rates of sediment building in the ocean are generally sluggish, often requiring thousands of years for substantial deposits to develop. Lithogenous sediment accumulates at a rapid rate, typically around 1 meter or more per thousand years for larger particles. Nevertheless, sedimentation rates in close proximity to the estuaries of substantial rivers with substantial water flow might be far greater in scale. Biogenous oozes accumulate at a pace of approximately 1 centimeter per millennium, whereas minute clay particles are deposited in the deep ocean at a rate of around 1 millimeter per millennium. According to section 12.4, manganese nodules accumulate at a very slow rate, increasing by only 0.001 mm per thousand years.

The thickness of marine sediments is greatest in close proximity to the continental margins, as seen in figure 12.1.1, reaching depths exceeding 10 kilometers. The reason for this is that the crust adjacent to passive continental margins is typically ancient, which allows for a prolonged time of sediment buildup. Additionally, there is a substantial influx of terrigenous material originating from the continents. In close proximity to mid-ocean ridge systems, where the creation of fresh oceanic crust takes place, the sediments are somewhat thinner due to the shorter period available for their accumulation on the recently created crust. As you distance yourself from the ridge spreading center, the sediments become increasingly thicker (see to section 4.5), growing by around 100-200 m of sediment for every 1000 km away from the ridge axis. The seafloor spreading pace of 20-40 km/million years corresponds to a sediment deposition rate of around 100-200 m during a period of 25-50 million years.

Figure 12.6.1 depicts the spatial arrangement of the primary categories of sediment found on the ocean floor. Cosmogenous sediments have the potential to be found in any area of the ocean, but their presence is so minimal that they are overshadowed by other types of sediment and hence do not prevail in any one location. Similarly, hydrogenous sediments may exhibit localized areas of high concentrations, but these zones are rather minuscule in terms of their global extent. In our discussion of global sediment patterns, we shall primarily disregard cosmogenous and hydrogenous deposits.

Coarse lithogenous/terrigenous sediments are dominant near the continental margins as runoff, river discharge, and other processes deposit vast amounts of these materials on the continental shelf (section 12.2). Much of this sediment remains on or near the shelf, while turbidity currents can transport material down the continental slope to the deep ocean floor. Lithogenous sediment is also common at the poles where thick ice cover can limit primary production, and glacial breakup deposits sediments along the ice edge. Coarse lithogenous sediments are less common in the central ocean, as these areas are too far from the sources for these sediments to accumulate. Very small clay particles are the exception, and as described below, they can accumulate in areas that other lithogenous sediment will not reach.

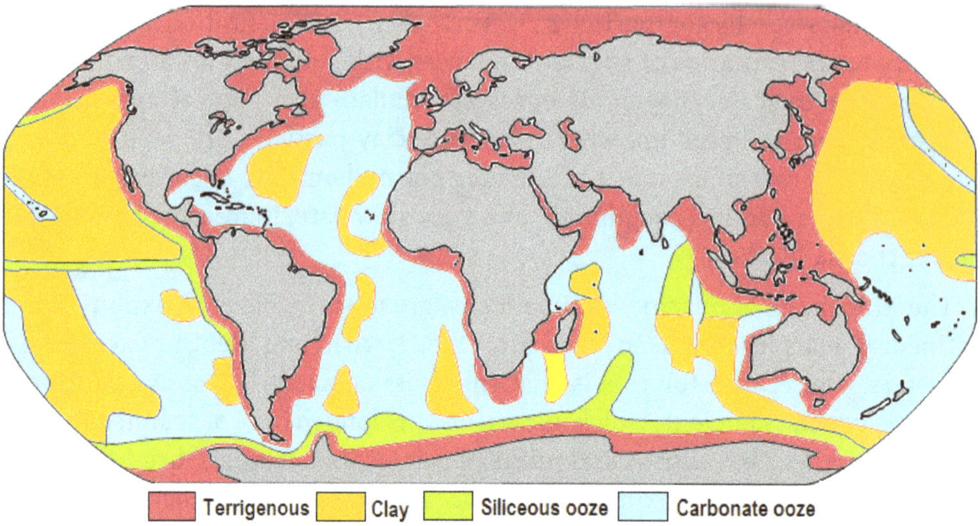

Figure 12.6.1 The distribution of sediment types on the seafloor. Within each colored area, the type of material shown is what dominates, although other materials are also likely to be present (Steven Earle, "Physical Geology").

The distribution of biogenous sediments depends on their rates of production, dissolution, and dilution by other sediments. We learned in section 7.4 that coastal areas display very high primary production, so we might expect to see abundant biogenous deposits in these regions. However, recall that sediment must be >30% biogenous to be considered a biogenous ooze, and even in productive coastal areas there is so much lithogenous input that it swamps the biogenous materials, and that 30% threshold is not reached. So coastal areas remain dominated by lithogenous sediment, and biogenous sediments will be more abundant in pelagic environments where there is little lithogenous input.

In order for biogenous sediments to accumulate their rate of production must be greater than the rate at which the tests dissolve. Silica is undersaturated throughout the ocean and will dissolve in seawater, but it dissolves more readily in warmer water and lower pressures; in other words, it dissolves faster near the surface than in deep water. Silica sediments will therefore only accumulate in cooler regions of high productivity where they accumulate faster than they dissolve. This includes upwelling regions near the equator and at high latitudes where there are abundant nutrients and cooler water. Oozes formed near the equatorial regions are usually dominated by radiolarians, while diatoms are more common in the polar oozes. Once the silica tests have settled on the bottom and are covered by subsequent layers, they are no longer subject to dissolution and the sediment will accumulate. Approximately 15% of the seafloor is covered by siliceous oozes.

Biogenous calcium carbonate sediments also require production to exceed dissolution for sediments to accumulate, but the processes involved are a little different than for silica. Calcium carbonate dissolves more readily in more acidic water. Cold seawater contains more dissolved CO_2 and is slightly more acidic than warmer water (section 5.5). Therefore calcium carbonate tests are more likely to dissolve in colder, deeper, polar water than in warmer, tropical, surface water. At the poles the water is uniformly cold, so calcium carbonate readily dissolves at all depths, and carbonate sediments do not accumulate. In temperate and tropical regions calcium carbonate dissolves more readily as it sinks into deeper water. The depth at which calcium carbonate dissolves as fast as it accumulates is called the **calcium carbonate compensation depth**, or **calcite compensation depth**, or simply the **CCD**. The **lysocline** represents the depths where the rate of calcium carbonate dissolution increases dramatically (similar to the thermocline and halocline). At depths shallower than the CCD carbonate accumulation will exceed the rate of dissolution, and carbonate sediments will be deposited. In areas deeper than the CCD, the rate of dissolution will exceed production, and no carbonate sediments can accumulate (Figure 12.6.2). The CCD is usually found at depths of 4 – 4.5 km, although it is much shallower at the poles where the surface water is cold. Thus calcareous oozes will mostly be found in tropical or temperate waters less than about 4 km deep, such as along the mid-ocean ridge systems and atop seamounts and plateaus. The CCD is deeper in the Atlantic than in the Pacific since the Pacific contains more CO_2, making the water more acidic and calcium carbonate more soluble. This, along with the fact that the Pacific is deeper, means that the Atlantic contains more calcareous sediment than the Pacific. All told, about 48% of the seafloor is dominated by calcareous oozes.

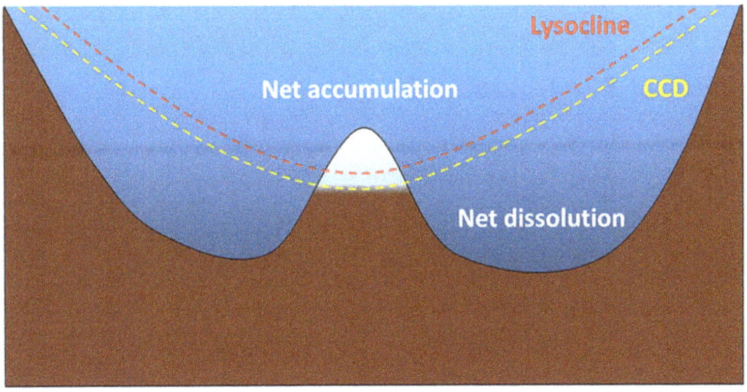

Figure 12.6.2 Calcareous sediment can only accumulate in depths shallower than the calcium carbonate compensation depth (CCD). Below the CCD, calcareous sediments dissolve and will not accumulate. The lysocline represents the depths where the rate of dissolution increases dramatically (PW).

A significant portion of the remaining deep ocean floor (about 38%) is primarily composed of abyssal clays. The presence of clay formation is primarily due to the absence of other sediment inputs, rather than an excessive supply of clay. The clay particles mostly originate from land, but due to their small size, they can be easily carried by wind and currents, allowing them to reach locations that are not accessible to other types of sediment. Clays are the predominant type of sediment in the middle North Pacific, as an illustration. This region is located at a considerable distance from the land, which prevents the transportation of coarse lithogenous sediment. Additionally, the location lacks the necessary conditions for the accumulation of biogenous tests, and its depth hinders the deposition of calcareous materials before they dissolve. Due to the gradual accumulation of clay particles, the deep ocean floor, which is predominantly clay, frequently contains hydrogenous deposits such as manganese nodules. If any alternative form of sediment were generated in this location, it would amass at a significantly faster rate, resulting in the burial of the nodules before they could undergo growth.

13 Chapter — COASTAL OCEANOGRAPHY: PROCESSES THAT OCCUR ALONG THE COAST

13.1 BEACHES

When most individuals envision coastal places, they typically visualize a beach, which is usually a conventional sandy beach made up of quartz sand grains (section 12.2). However, beaches consist of the predominant types of sediments found in the surrounding region. For instance, many regions in Hawaii and Iceland are renowned for their black sand beaches, composed of weathered basalt and other volcanic substances. The stunning tropical white sand beaches depicted in tourist advertisements primarily consist of the pulverized calcium carbonate remnants of coral skeletons, a significant portion of which has been ingested and expelled by fish before we joyfully walk barefoot on it. Some beaches may not have any sand and instead have a predominant presence of small shells, larger rocks, or pebbles (Figure 13.1.1).

The coastline is partitioned into many zones (Figure 13.1.2). The backshore refers to the portion of the beach that lies above the high tide line. It is typically only covered by water during exceptional wave events, such as hurricanes. The foreshore is the area that is located between the high tide and low tide lines. It is underwater during high tide and becomes visible during low tide. The nearshore zone encompasses the area between the low tide line to the point when wave action is no longer affected by the seabed, specifically where the depth surpasses the wave base (section 10.1). The offshore zone refers to the areas that lie outside the nearshore region.

The section of the beach that is located above the high tide line is referred to as the berm. Typically, the berm is dry and has a rather even surface. The berm typically terminates with a berm crest or berm scarp, which is a more inclined wall formed by the force of waves and descends towards the beach. The foreshore is alternatively referred to as the beach face, the intertidal or littoral zone, and, in cases when the region is relatively level, the low tide terrace. Frequently, there are elongated sandbars and depressions running parallel to the beach, located just off the shore. Longshore bars are sand deposits formed by the combined forces of waves and longshore currents (see 13.2). The reduction in water depth across longshore bars frequently leads to the initiation of wave breaking well in advance of reaching the shoreline (section 10.3).

Figure 13.1.1 Various beach substrates. Clockwise from top left: Punaluu Black Sand Beach, Hawaii, USA (Diego Delso [CC BY-SA 3.0], via Wikimedia Commons); Shell Beach, Shark Bay, Western Australia (Brian W. Schaller (Own work) [CC BY-NC-SA 3.0], via Wikimedia Commons); white coral sand beach in the Maldives (http://www.elitedivingagency.com/, [CC BY-SA 3.0] via Wikimedia Commons); rocky beach at Killbear Provincial Park in Ontario, Canada (John Vetterli (originally posted to Flickr as Beach) [CC BY-SA 2.0], via Wikimedia Commons).

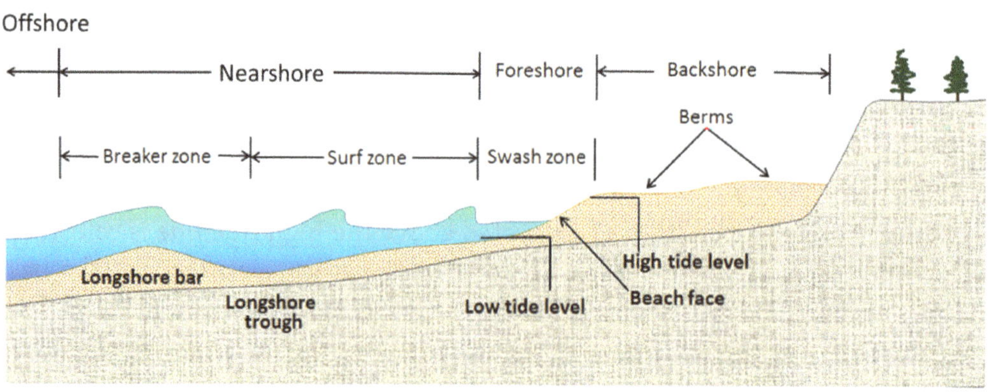

Figure 13.1.2 The zones of a typical beach (Modified by PW from Steven Earle, "Physical Geology").

Wave motion disperses the sand or other particles that constitute the beach. The water that flows onto a beach as a result of incoming waves is referred to as swash. This water also carries sand grains that are suspended in it, which can then be deposited on the beach. A portion of the swash infiltrates the sand, while the remaining water retreats as backwash when the wave recedes. Backwash transports sand from the beach and deposits it back into the ocean. Consequently, sand will either accumulate or erode, depending on the prevailing mechanism. When wave movement is gentle, a significant amount of water is absorbed by the sand, resulting in the dominance of swash. When the waves are stronger, the beach gets filled with water, making it unable to absorb as much water. As a result, the backwash becomes the main force. As a consequence of stormier conditions at sea, waves become more intense throughout the winter, resulting in seasonal changes in beach structure. throughout this time, backwash becomes more prominent, causing the removal of sand from the beach and its deposition offshore in the form of longshore bars. During the summer, the waves are less forceful, with swash being the dominant motion. As a result, sand from the longshore bar is carried and placed on the shore, resulting in the formation of a broader beach made of sand (Figure 13.1.3).

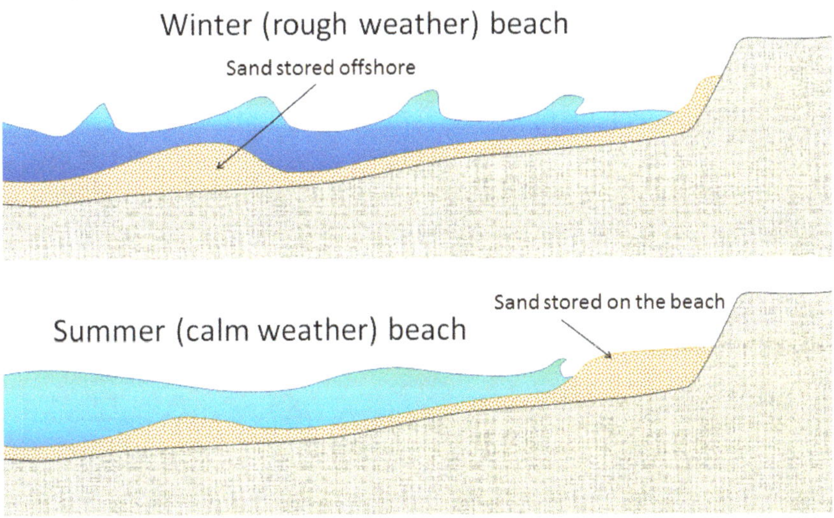

Figure 13.1.3 The differences between summer and winter on beaches in areas where the winter conditions are rougher and waves have a shorter wavelength but higher energy. In winter, sand from the beach is stored offshore (Steven Earle, "Physical Geology").

13.2 LONGSHORE TRANSPORT

*Modified from "Physical Geology" by Steven Earle**

We learned in section 10.3 that refraction causes waves to approach parallel to shore. However, most waves still reach the shore at a small angle, and as each one arrives, it pushes water along the shore, creating what is known as a **longshore current** within the surf zone (the areas where waves are breaking) (Figure 13.2.1). Longshore currents can move up to 4 km/hr, strong enough to carry people with them, as everyone knows who has been swimming in the ocean only to look up and see that they have been carried far down the beach from their towel!

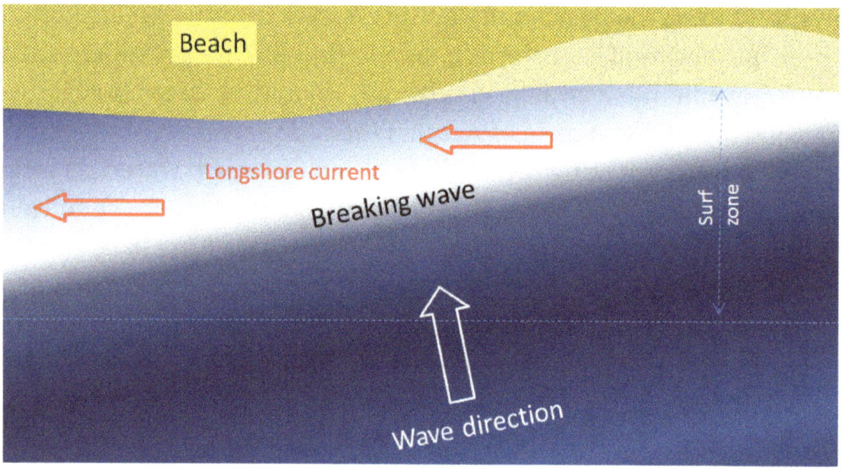

Figure 13.2.1 Longshore currents are caused by waves approaching shore at a small angle, moving water parallel to the shore (Steven Earle, "Physical Geology").

Another important effect of waves reaching the shore at an angle is that when they wash up onto the beach, they do so at an angle, but when that same wave water washes back down the beach, it moves straight down the slope of the beach (Figure 13.2.2). The upward-moving water, known as the swash, pushes sediment particles along the beach, while the downward-moving water, the backwash, brings them straight back. With every wave that washes up and then down the beach, particles of sediment are moved along the beach in a zigzag pattern.

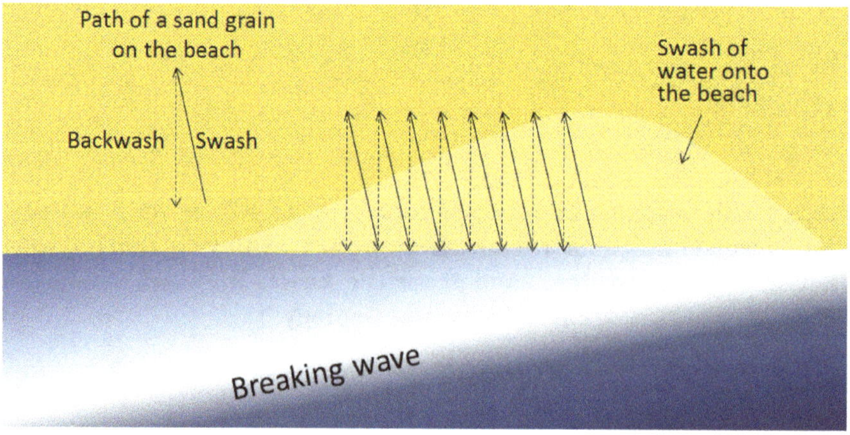

Figure 13.2.2 The zigzag pattern of sediment movement along a beach creating longshore transport. In this figure the longshore transport moves particles to the left (Steven Earle, "Physical Geology").

A **rip current** (often incorrectly called a "rip tide"; they are not really related to tides) is another type of current that develops in the nearshore area, and has the effect of returning water that has been pushed up to the shore by incoming waves or accumulated through longshore currents, particularly converging longshore currents. Rip currents often occur where there is a channel between sandbars that makes it easier for the retreating water to escape. As shown in Figure 13.2.3, rip currents flow straight out from the shore, and because the water is directed through a narrow space, the current can be very strong. The currents lose strength quickly just outside of the surf zone, but they can be dangerous to swimmers who get caught in them and are pulled away from shore. Swimmers caught in a rip current should not try to swim directly back to shore, as it is difficult to fight the current and the swimmer can quickly tire. Instead, swim parallel to the beach for a short distance until you are outside of the rip current, and then you can easily swim to shore.

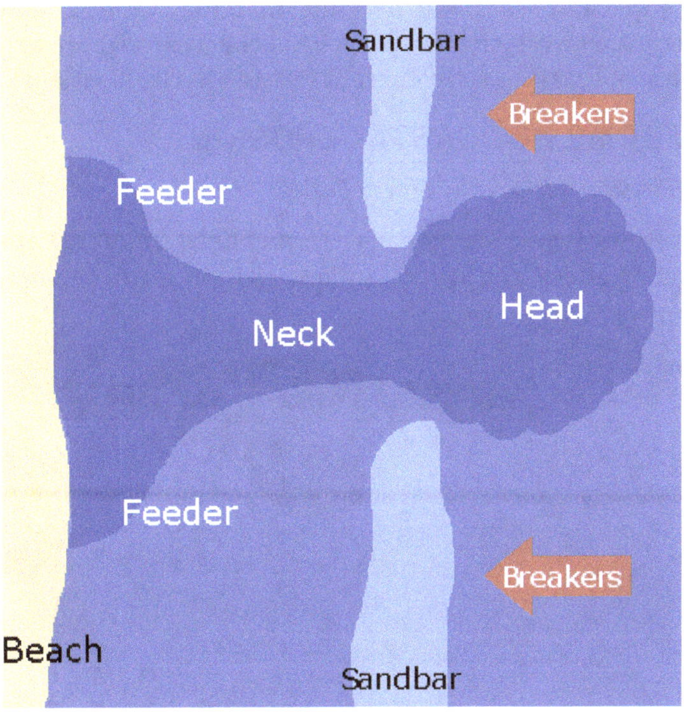

Figure 13.2.3 Creation of a rip current from wave action and longshore transport. Water accumulates on the beach, and then rushes out to sea through a narrow channel, creating a strong current (National Weather Service, Wilmington, NC (NOAA) [Public domain], via Wikimedia Commons).

Rip currents are visible in Figure 13.2.4, a beach at Tunquen in Chile near Valparaiso. As is evident from the photo, the rips correspond with embayments in the beach profile. Three of them are indicated with arrows, but it appears that there may be several others farther along the beach.

Figure 13.2.4 Rip currents along a beach in Chile, indicated by the arrows. Longshore currents converging in a curved beach have nowhere to go but straight back out to sea, creating a rip current (Steven Earle, "Physical Geology").

13.3 LANDFORMS OF COASTAL EROSION

*Modified from "Physical Geology" by Steven Earle**

Large waves crashing onto a shore bring a tremendous amount of energy that has a significant eroding effect, and several unique erosion features commonly form on rocky shores with strong waves.

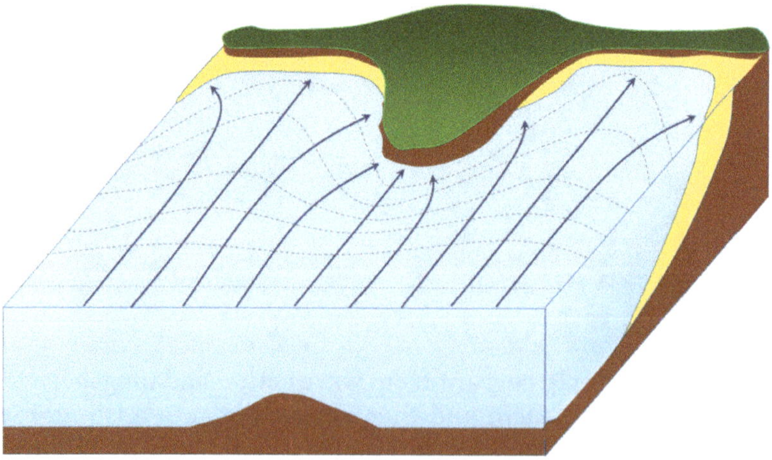

Figure 13.3.1 The approach of waves (blue lines) towards a coastal headland. The blue arrows represent wave energy; most of that energy is focused on the headlands, causing greatest erosion in this area (PW).

When waves approach an irregular shore, they are slowed down to varying degrees, depending on differences in the water depth, and as they slow, they are bent or refracted (section 10.3). In Figure 13.3.1, wave energy is represented by the blue arrows. That energy is evenly spaced out in the deep water, but because of refraction, the energy of the waves is being focused on the **headlands.** On irregular coasts, the headlands receive much more wave energy than the intervening bays, and thus they are more strongly eroded. The result of this is **coastal straightening,** where an irregular coast will eventually become straightened, although that process may take millions of years.

Wave erosion is greatest in the surf zone, where the wave base is impinging strongly on the seafloor and where the waves are breaking. The result is that the substrate in the surf zone is typically eroded to a flat surface known as a **wave-cut platform** (or **wave-cut terrace**) (Figure 13.3.2). A wave-cut platform extends across the intertidal zone.

Figure 13.3.2 A wave-cut platform in bedded sedimentary rock on Gabriola Island, B.C. The wave-eroded surface is submerged at high tide (Steven Earle, "Physical Geology").

Arches and sea caves are created through the process of eroding rock that is comparatively less resistant to weathering. Wave action and powerful longshore currents have the ability to sculpt a cave into a headland. If the erosion continues until it penetrates completely, it transforms into an arch. In the event that a cavity in the ceiling of a cave forms, it can give rise to a blowhole, which forcefully propels water into the atmosphere upon the impact of waves inside the cave. Figure 13.3.3 displays an arch located in the Barachois River region of western Newfoundland, Canada. Initially, this characteristic originated as a maritime cavern, and then, due to erosion from opposing directions, transformed into an arch. In the winter of 2012-2013, the arch fell, resulting in a little stack remaining at the end of the point.

Figure 13.3.3 Top: An arch in tilted sedimentary rock at the mouth of the Barachois River, Newfoundland, July 2012. Bottom: The same location in June 2013. The arch has collapsed and a small stack remains (Photo: Dr. David Murphy, used with permission in Steven Earle, "Physical Geology").

A sea stack is the term used to describe the tower of rock that remains after an arch has fallen (Figure 13.3.4). Sea stacks can also develop during the creation of wave-cut platforms or other geological formations, when durable rock that is not fully eroded remains intact and forms the stack.

Figure 13.3.4 A sea stack, likely created from the collapse of a sea arch (Doug Lee [CC BY-SA 2.0], via Wikimedia Commons).

Figure 13.3.5 provides an overview of the process by which an irregular coastline is transformed into a straightened coastline with sea cliffs (escarpments worn by waves) and the remaining formations of stacks, arches, and wave-cut platforms. The latter phases of this process entail the ongoing inward erosion of the coastal cliffs and the total erosion of the rock formations and platforms created by wave action, resulting in the formation of a continuous and predominantly linear sandy shoreline.

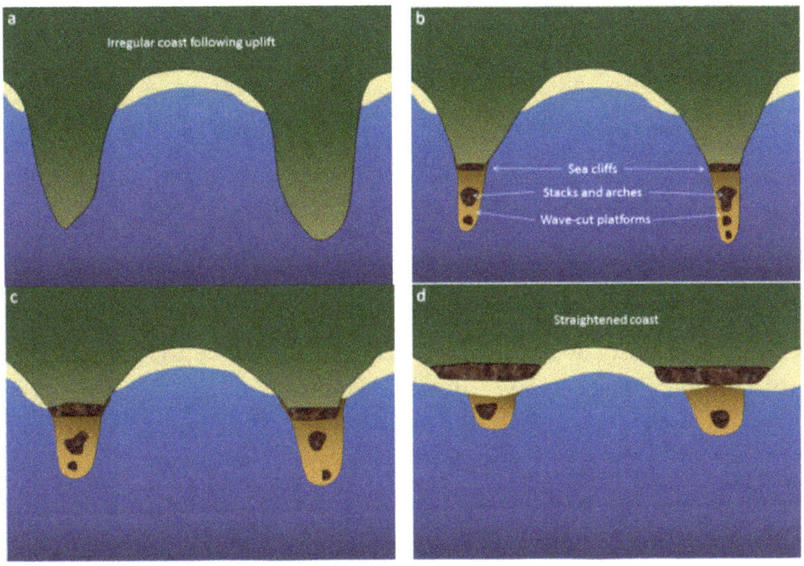

Figure 13.3.5 Evolution of a straightened coast through the erosion to stacks and arches, sea cliffs, and wave-cut platforms (Steven Earle, "Physical Geology").

13.4 LANDFORMS OF COASTAL DEPOSITION

*Modified from "Physical Geology" by Steven Earle**

Coastal areas can be categorized into two types: those that are mostly affected by erosion, such as the Pacific coast of North America, and those that are primarily affected by deposition, such as the Atlantic and Caribbean coasts of the United States. However, in most coastal areas, both the process of sediment deposition and erosion occur to variable extents, albeit in distinct locations. Deposition-dominant coasts experience erosion in certain regions while sediment is being deposited in other areas.

In coastal areas where depositional processes prevail, the majority of sediment being deposited usually originates from major rivers. A significant portion of the sediment is rapidly deposited around the river's mouth, resulting in the formation of expansive deltas with a fan-like appearance. A clear illustration may be seen when the Mississippi River converges with the Gulf of Mexico in New Orleans; another instance is the Yellow (Huang He) River in China (Figure 13.4.1).

Figure 13.4.1 The Yellow River delta in China, created by one of the most sediment-laden rivers on Earth (NASA [Public domain], via Wikimedia Commons).

The evolution of sandy depositional features on sea coasts is primarily influenced by waves and currents, especially longshore currents. As sediment is transported along a shore, either it is deposited on beaches, or it creates other depositional features. A **spit**, for example is an elongated sandy deposit that extends out into open water in the direction of a longshore current (Figure 13.4.2).

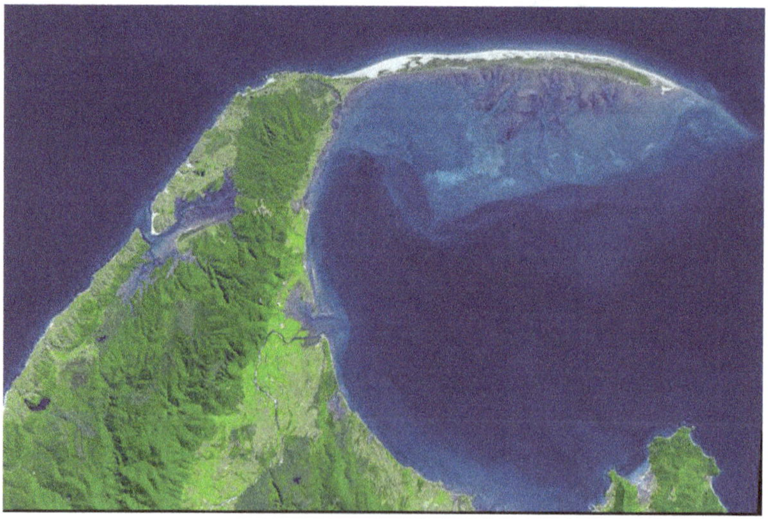

Figure 13.4.2 Farewell Spit, on the northern shore of New Zealand's South Island (By NASA/GSFC/METI/ERSDAC/JAROS, and U.S./Japan ASTER Science Team (NASA's Earth Observatory) [Public domain], via Wikimedia Commons).

A baymouth bar refers to a spit that stretches across a bay, thus sealing it off or nearly closing it off. Streams often flow into most bays, and as the water needs to exit, it is uncommon for a baymouth bar to fully block the entrance to a bay.

Figure 13.4.3 Left: Illustration of a baymouth bar and tombolo (Steven Earle, "Physical Geology"). Right: A baymouth bar at the mouth of the Klamath River in northern California (Linda Tanner, https://www.flickr.com/photos/goingslo/5827465324, Creative Commons CC-BY 2.0).

Tombolos are common where islands are abundant, and they typically form where there is a wave shadow behind a nearshore island (Figure 13.4.4). This becomes an area with reduced energy, and so the longshore current slows and sediments accumulate. Eventually enough sediments accumulate to connect the island to the mainland with a tombolo (Figure 13.4.5).

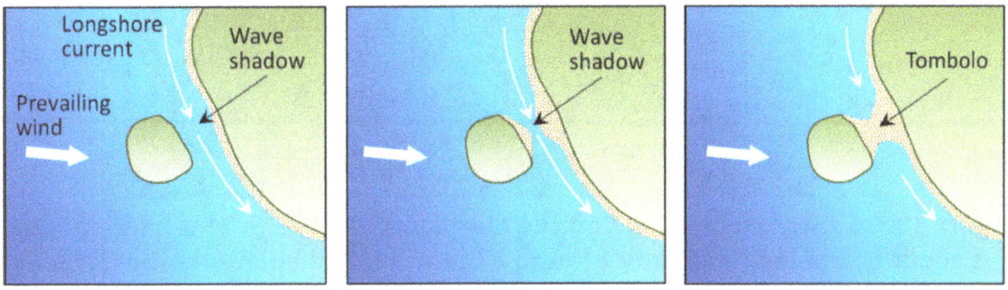

Figure 13.4.4 Formation of a tombolo. In the wind shadow of an island, there is little wave action, so the sediment moved by longshore transport gets deposited, eventually linking the island to the mainland (Steven Earle, "Physical Geology").

Figure 13.4.5 A stack (with a wave-cut platform) connected to the mainland by a tombolo, Leboeuf Bay, Gabriola Island, British Columbia (Steven Earle, "Physical Geology").

Barrier islands often develop in regions with ample coastal sediments and minimal coastal elevation changes due to limited or absent recent uplift (as depicted in Figure 13.4.6). Barrier islands are elongated landforms consisting of sand that develop parallel to the mainland, often extending several kilometers in width and hundreds of kilometers in length. They are prevalent in the U.S. Gulf Coast spanning from Texas to Florida, and along the U.S. Atlantic Coast extending from Florida to Massachusetts. Islands often emerge when silt is transported from land to the sea by river discharge, and then wave action redistributes the sediment towards the shore.

Figure 13.4.6 Assateague Island on the Maryland coast, U.S. This barrier island is about 60 km long and only 1 km to 2 km wide. The open Atlantic Ocean is to the right and the lagoon is to the left. This part of Assateague Island has recently been eroded by a tropical storm, which pushed massive amounts of sand into the lagoon. (http://soundwaves.usgs.gov/2014/04/images/DelmarvaAssateague_aerial_ViewCV.jpg).

Mature barrier islands contain a number of ecological zones. Beginning on the ocean side of the island there is a beach, consisting of the zones we discussed in section 13.1. Behind the beach lie dunes that are built up by sand transported by wind. The ocean side of the dunes are home to grasses and other plants which help stabilize the sand from erosion, and also help slow down the wind to allow sand to settle and accumulate. Beyond the dunes lies a more heavily vegetated barrier flat, covered by larger shrubs and trees that are tolerant to the high winds and salty conditions. As the land slopes down on the side of the island facing the mainland, the low-lying areas transition into a salt marsh or mud flat habitat, which is protected from wave action, but is influenced by tidal changes. The mud flats are colonized by grasses, which slow down the movement of water and lead to increased sediment deposition, building up the land in the marsh. Different species of grasses eventually dominate the different elevations of the salt marsh, depending on their tolerance for submersion in seawater. These salt marshes are very important habitats for many invertebrates, birds, and juvenile fish. Between the island and the mainland lies a lagoon, which usually contains brackish water from the mixing of fresh water runoff from the land and the seawater within a somewhat enclosed space. Barrier islands, although attractive locations for beach houses, are not permanent structures, and people should be wary of building on them. Over time, the erosion on the seaward side of the island, and the expansion of the marsh on the landward side, causes the island to slowly move towards the mainland, eventually closing off the lagoon. Maintaining dune grasses is one way to slow this movement, and as we will see in the next section, people have developed a number of other strategies to try to curtail the natural erosion of beaches.

13.5 HUMAN INTERFERENCE WITH SHORELINES

The continued erosion and deposition of coastal sediments is a natural process, with features forming and disappearing as sea level and other conditions change. However, we have also come to enjoy and rely on many of these beaches and other coastal features for commerce, recreation, and living space. So from our perspective, we often see the transient nature of the coast as a threat to our activities, and as a result we have developed a number of ways to try and influence the erosion process, usually through **hard stabilization**, or the building of structures to stop the flow of sand. While some of these efforts have been successful, many others have actually exacerbated the problem, as we will see below.

Figure 13.5.1 Left: A groin on Bolinas Beach, California. Longshore transport would be moving from right to left, accumulating sand on the right of the groin (Beatrice Murch, https://www.flickr.com/photos/blmurch/952523971, CC BY-SA 2.0). Right: A groin field. Longshore transport would move from the bottom of the picture towards the top (https://pxhere.com/en/photo/1048378, Public Domain, CC0).

Jetties are like longer groins, often built to protect the mouths of harbors to prevent them from filling with sand. Because they are longer they can trap more sand than groins, and they also can contribute to increased erosion on the downstream side (Figure 13.5.2). If too much sand accumulates upstream of the jetty it can spread past the jetty and into the mouth of the harbor, in which case the jetty may need to be extended.

Figure 13.5.2 The jetties protecting Santa Cruz harbor, California. Sand has accumulated on the left (north) side of the harbor, and the beach is eroded on the right (south) side (Google Maps, Map Data: Google).

Breakwaters are walls that are usually built parallel to the shore. Their purpose is not necessarily to interfere with sediment transport, but instead to protect the areas behind them from heavy wave action, so they are often deployed at the mouth of a harbor, or to protect the boats in a mooring field. But breakwaters do have an unintended impact on sediment distribution. Longshore transport continues to move sand along the beach, but once it gets behind the breakwater the lack of wave action interrupts the flow, and the sand settles and accumulates. The beach grows behind a breakwater, until eventually they may become connected (Figure 13.5.3). With the longshore transport interrupted, increased erosion can occur downstream of the breakwater.

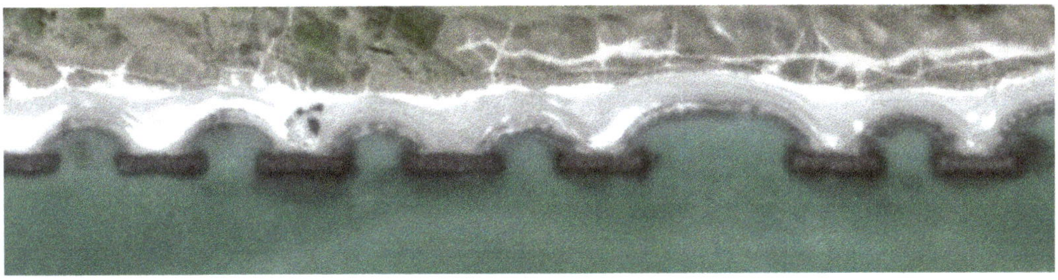

Figure 13.5.3 A series of breakwaters along the coast in Skagen, Denmark. In this case, sand has accumulated to the point that they now connect the breakwaters to land (Google Maps; Imagery: DigitalGlobe, Aerodata,International Surveys, TerraMetrics, Data SIO, NOAA, US Navy, NGA, GEBCO, Map Data: Google).

Santa Monica Pier

The beach around the Santa Monica Pier in California provides a good example of the effects of breakwaters on a sandy shore. A breakwater was constructed in the early 1930s to protect the pier and the boats that moored near it. Following the construction, the once-straight beach became much wider behind the breakwater as sand accumulated in the absence of strong wave action (*below left*). Now that the breakwater is no longer in place, the bulge in the shoreline is gone, and the beach is much straighter once again (*below right*).

Left; aerial image of Santa Monica Pier and breakwater from 1936 (Courtesy of Santa Monica Public Library Image Archives/ Spence Air Photos). Right; Santa Monica Pier in 2011 (© JCS, CC BY-SA 3.0, via Wikimedia Commons).

Seawalls are constructed at the top of the surf zone, where the waves crash against the shore. The walls are designed as a barrier between the waves and the shore, to prevent the land from being eroded (Figure 13.5.4). They are often utilized in beachfront property to prevent the ground under a home from being undermined by the waves. However, as with the other forms of hard stabilization we have discussed, seawalls are not without their own environmental consequences. The sudden release of wave energy on a seawall can create turbulence, which undermines the sediment at the base of the wall and causes it to erode. Furthermore, on a softer, natural coastline some of the wave energy is absorbed or dissipated, but with a hard seawall most of the wave energy is reflected, leading to stronger longshore currents and faster erosion. In many places where seawalls have been built the beaches are getting steeper, and erosion rates have increased, with the potential for seawalls to collapse along with whatever they are supporting. Because of this, some coastal communities are phasing out seawall construction to try to return to more natural beach fronts.

Figure 13.5.4 A sea wall at Horsey Gap, UK (Evelyn Simak [CC BY-SA 2.0], via Wikimedia Commons).

13.6 ESTUARIES

Estuaries are partially enclosed bodies of water where the salt water is diluted by fresh water input from land, creating **brackish** water with a salinity somewhere between fresh water and normal seawater. Estuaries include many bays, inlets, and sounds, and are often subject to large temperature and salinity variations due to their enclosed nature and smaller size compared to the open ocean.

Estuaries can be classified geologically into four basic categories based on their method of origin. In all cases they are a result of rising sea level over the last 18,000 years, beginning with the end of the last ice age; a period that has seen a rise of about 130 m. The rise in sea level has flooded coastal areas that were previously above water, and prevented the estuaries from being filled in by all of the sediments that have been emptied into them.

The first type is a **coastal plain estuary**, or **drowned river valley**. These estuaries are formed as sea level rises and floods an existing river valley, mixing salt and fresh water to create the brackish conditions where the river meets the sea. These types of estuaries are common along the east coast of the United States, including major bodies such as the Chesapeake Bay, Delaware Bay, and Narragansett Bay (Figure 13.6.1). Coastal plain estuaries are usually shallow, and since there is a lot of sediment input from the rivers, there are often a number of depositional features associated with them such as spits and barrier islands.

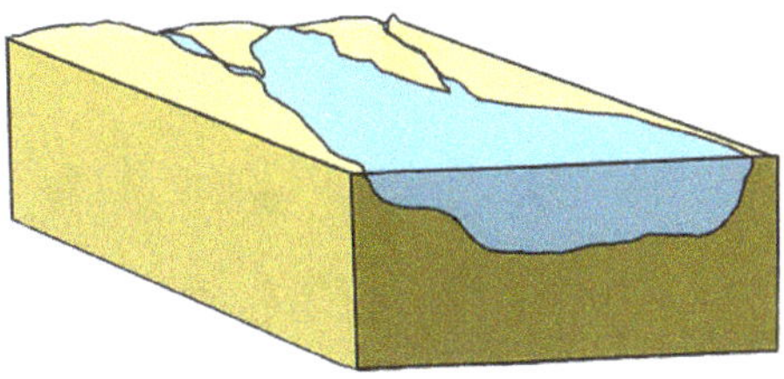

Figure 13.6.1 A coastal plain estuary. Sea level has risen and flooded what was once a river valley. The satellite image shows Chesapeake Bay and Delaware Bay, two coastal plain estuaries (left: JR, right: NASA, Public Domain via Wikimedia Commons).

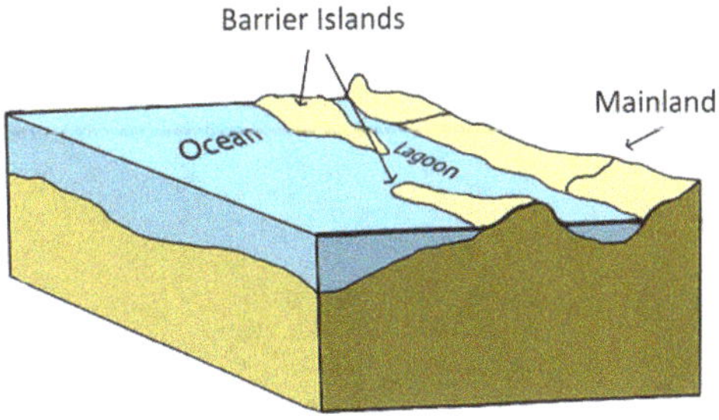

Figure 13.6.2 A bar-built estuary. Sand bars and barrier islands have partially isolated a lagoon from the rest of the ocean. Freshwater input into the lagoon from the mainland creates brackish conditions in the estuary. At right is a satellite image of Pamlico Sound, North Carolina, a bar-built estuary surrounded by spits and barrier islands (left: JR, right: NASA, Public Domain via Wikimedia Commons).

The presence of sand bars, spits, and barrier islands can lead to **bar-built estuaries**, where a barrier is created between the mainland and the ocean. The water that remains inside the sand bar is cut off from complete mixing with the ocean, and receives freshwater input from the mainland, creating estuarine conditions (Figure 13.6.2).

Fjords are estuaries formed in deep, U-shaped basins that were carved out by advancing glaciers. When the glaciers melted and retreated, sea level rose and filled these troughs, creating deep, steep-walled fjords (Figure 13.6.3). Fjords are common in Norway, Alaska, Canada, and New Zealand, where there are mountainous coastlines once covered by glaciers.

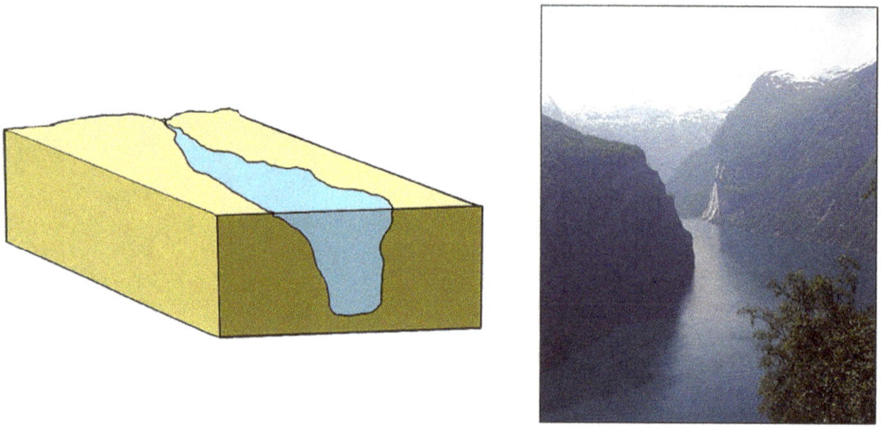

Figure 13.6.3 A fjord is a deep estuary that was carved out by glacial movements. At right is Geirangerfjord, Norway (left: JR, right: Fgmedia, [CC-BY-SA-3.0], via Wikimedia Commons).

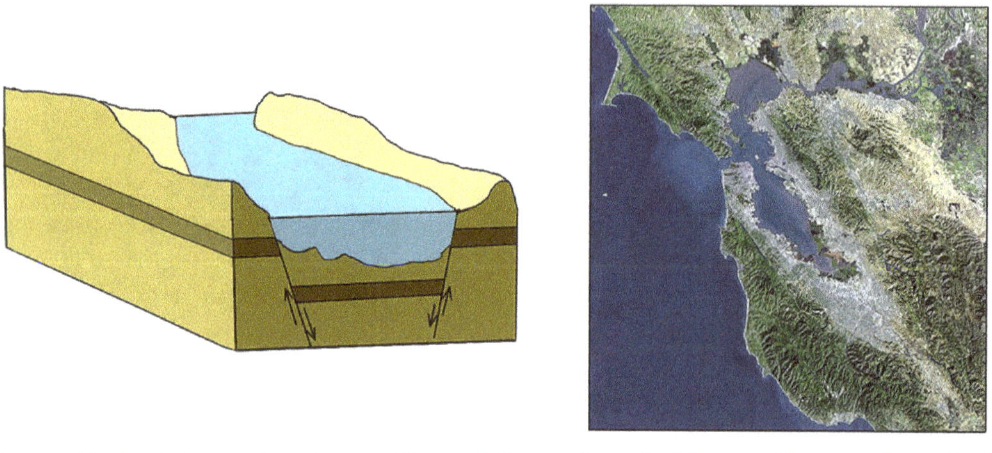

Figure 13.6.4 A tectonic estuary, formed from the subsidence of crust along fault lines, and the subsequent filling by seawater. San Francisco Bay is a tectonic estuary, shown at right (left: JR, right: USGS, Public Domain, via Wikimedia Commons).

Tectonic estuaries are the result of tectonic movements, where faulting causes some sections of the crust to subside, and those lower elevation sections then get flooded with seawater. San Francisco Bay is an example of a tectonic estuary (Figure 13.6.4).

Estuaries are also classified based on their salinity and mixing patterns. The amount of mixing of fresh and salt water in an estuary depends on the rate at which fresh water enters the head of the estuary from river input, and the amount of seawater that enters the estuary mouth as a result of tidal movements. The input of fresh water is reflected in the **flushing time** of the estuary. This refers to the time it would take for the in-flowing fresh water to completely replace all the fresh water currently in the estuary. Seawater input is measured by the **tidal volume**, or **tidal prism**, which is the average volume of sea water entering and leaving the estuary during each tidal cycle. In other words, it is the volume difference between high and low tides. The interaction between the flushing time, tidal volume, and the shape of the estuary will determine the extent and type of water mixing within the estuary.

In a **vertically mixed**, or **well-mixed** estuary there is complete mixing of fresh and salt water from the surface to the bottom. In a particular location the salinity is constant at all depths, but across the estuary the salinity is lowest at the head where the fresh water enters, and is highest at the mouth, where the seawater comes in. This type of salinity profile usually occurs in shallower estuaries, where the shallow depths allow complete mixing from the surface to the bottom.

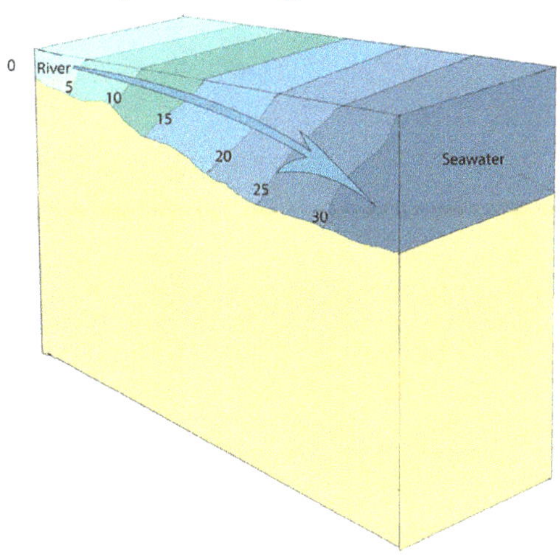

Figure 13.6.5 A well-mixed estuary. The shallow basin allows nearly complete mixing or fresh and seawater from top to bottom. Salinities are in ppt (JR).

Slightly stratified or **partially mixed** estuaries have similar salinity profiles to vertically mixed estuaries, where salinity increases from the head to the mouth, but there is also a slight increase in salinity with depth at any point. This usually occurs in deeper estuaries than those that are well-mixed, where waves and currents mix the surface water, but the mixing may not extend all the way to the bottom.

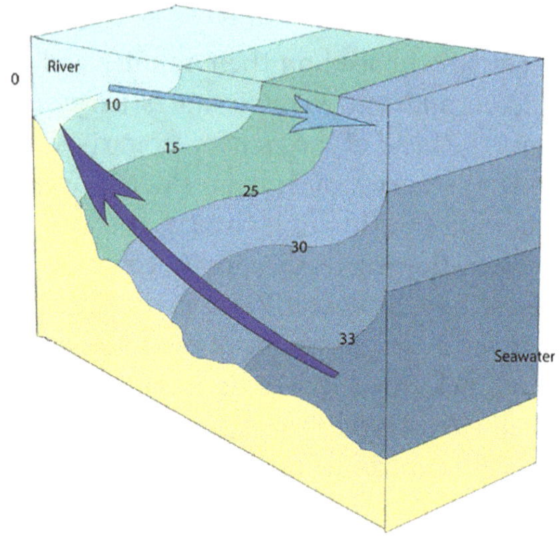

Figure 13.6.6 A slightly-stratified estuary. Generally deeper than a well-mixed estuary, the inflow of seawater (dark blue arrow) and fresh water (light blue arrow) create an estuary where salinity increases with depth, and at the surface when moving from the head to the mouth of the estuary. Salinities are in ppt (JR).

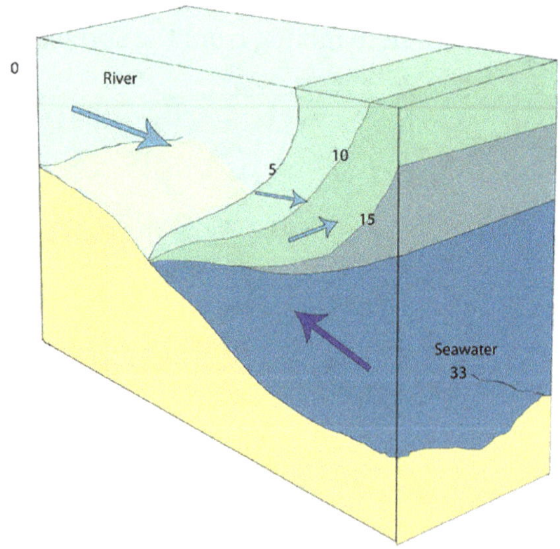

Figure 13.6.7 A salt wedge estuary. Strong river outflow (light blue arrows) creates a layer of mostly fresh water that sits on top of a wedge of encroaching seawater along the bottom (dark blue arrow). Salinities are in ppt (JR).

A **salt wedge** estuary occurs where the outflow of fresh water is strong enough to prevent the denser ocean water to enter through the surface, and where the estuary is deep enough that surface waves and turbulence have little mixing effect on the deeper water. Fresh water flows out along surface, salt water flows in at depth, creating a wedge shaped lens of seawater moving along the bottom. The surface water may remain mostly fresh throughout the estuary if there is no mixing, or it can become brackish depending on the level of mixing that occurs.

Highly stratified profiles are found in very deep estuaries, such as in fjords. Because of the depth, mixing of fresh and salt water only occurs near the surface, so in the upper layers salinity increases from the head to the mouth, but the deeper water is of standard ocean salinity.

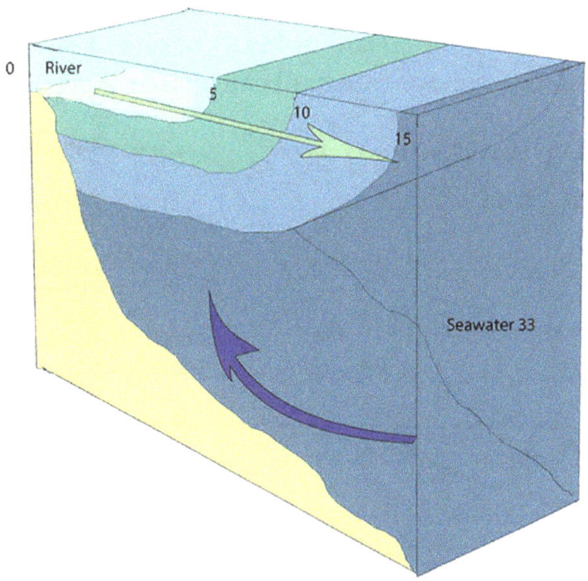

Figure 13.6.8 A highly stratified estuary. Strong river outflow and a deep basin prevent mixing between surface and bottom water, creating an estuary that is vertically stratified. Salinities are represented in ppt (JR).

Estuaries are very important commercially, as they are home to the majority of the world's metropolitan areas, they serve as ports for industrial activity, and a large percentage of the world's population lives near estuaries. Estuaries are also very important biologically, especially in their role as the breeding grounds for many species of fish, birds, and invertebrates.

13.7 SEA LEVEL CHANGE

*Modified from "Physical Geology" by Steven Earle**

Sea level fluctuation has been a prominent characteristic of Earth for billions of years, and it significantly affects coastal processes, estuaries, as well as erosion and deposition of landforms. Sea level change can occur through two primary mechanisms: eustatic and isostatic.

Eustatic sea level changes refer to global fluctuations in sea level that are caused by variations in the ocean's water volume. These variations can arise from alterations in the mass of ice on land, expansion of water due to temperature changes, or modifications in the topography of the ocean floor induced by plate tectonics. Seafloor spreading, as an illustration, causes the expansion of an ocean basin, hence altering its volume and impacting sea level.

During the last 20,000 years, the melting of glaciers has caused a rise in eustatic sea level of around 125 m. The majority of these events occurred between 15,000 and 7,500 years ago, which coincided with the significant melting period of the North American and Eurasian Ice Sheets (as shown in Figure 13.7.1). Approximately 7,500 years ago, there was a significant drop in the rate of glacier melting and sea level rise. Since then, the average rate has been approximately 0.7 mm/year.

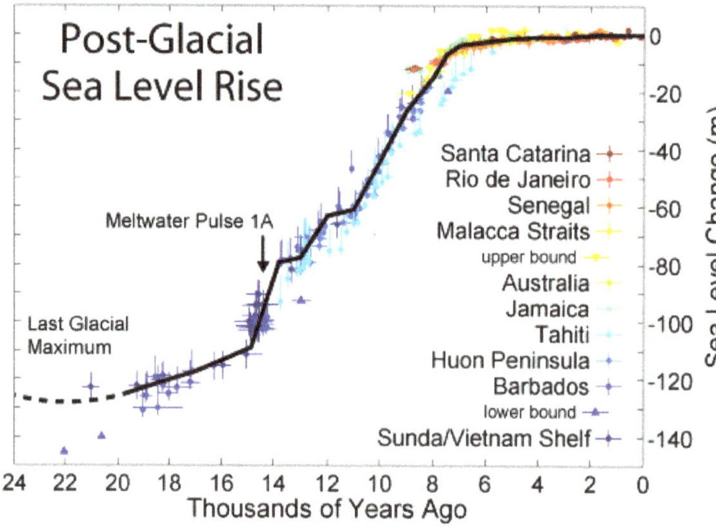

Figure 13.7.1 Sea level rise resulting from the melting of glacial ice over the past 24,000 years (Robert A. Rhode, CC BY-SA 3.0, via Wikimedia Commons).

The rise in sea levels began to accelerate around 1870 as a result of human-induced climate change. Since then, the average rate has been 1.1 millimeters per year, with a moderate upward trend. From 1992 onwards, the average rate has always been 3.2 millimeters per year, as shown in Figure 13.7.2. The primary cause of this phenomenon is the escalating global temperature, which leads to amplified glacier melting (section 14.3). However, a significant portion of the effect can be attributed to the expansion of water caused by thermal changes. As the temperature of water increases, the molecules inside it acquire greater kinetic energy, causing them to move with increased speed and distance from each other. Consequently, the volume occupied by the same quantity of water expands. Therefore, even in the absence of additional water intake from melting ice, the increase in ocean temperatures will result in a rise in sea level.

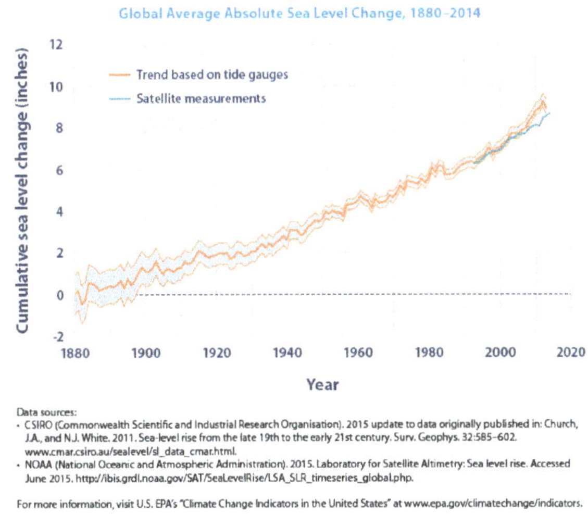

Figure 13.7.2 Average absolute sea level change, which refers to the height of the ocean surface, regardless of whether nearby land is rising or falling. Satellite data are based solely on measured sea level, while the long-term tide gauge data include a small correction factor because the size and shape of the oceans are changing slowly over time (US EPA [Public domain], via Wikimedia Commons).

Isostatic sea level fluctuations refer to localized variations in sea level resulting from the sinking or rising of the Earth's crust, which can be attributed to either changes in the quantity of ice on land or the formation and erosion of mountains. During the height of the last glaciation, a significant portion of Canada and several areas in the northern United States were enveloped by extensive ice sheets. After the ice melted, there was a phenomenon called isostatic rebound, where the continental crust rose in numerous regions. The rebound of the Laurentide Ice Sheet varies from a few hundred meters in the center region around Hudson Bay to 100 to 200 meters in locations like Vancouver Island and the mainland coast of British Columbia. To clarify, while the global sea level was approximately 130 meters lower during the last glaciation, the areas covered by glaciers were pushed down by at least the same amount in most locations, and even more in areas where the ice was thickest. Tectonic processes, such as the elevation of Earth's crust, can also result in specific alterations in the level of the sea.

14 Chapter

ICE: ICE FORMATION, TYPES OF ICE THAT ARE FOUND IN THE OCEAN, AND THE IMPACTS OF CLIMATE CHANGE

Approximately 6% of the Earth's surface is covered by ice, which contributes to the creation of stunning and magnificent oceanic features. However, the impact of ice on global weather and climate is often underestimated. As an illustration:

- The presence of ice on the Earth's surface influences how much solar radiation is absorbed by changing the reflectance (albedo).
- Ice acts as insulation, preventing significant changes in polar sea temperatures by blocking heat exchange between the ocean and the atmosphere. A one-meter layer of ice significantly decreases the rate of heat transfer between the Earth and the atmosphere by a factor of 100.
- Seasonal fluctuations in temperature, caused by the freezing and thawing of ice, help regulate ocean temperatures by utilizing the latent heat of freezing and fusion.Ice plays a role in several ways in causing the uneven heating of Earth's surface, which forms the foundation for global wind and climate patterns.

(By NASA / James Yungel (Iceberg in sea ice) [Public domain], via Wikimedia Commons).

In this section, we will specifically analyze the process of ice production, the many types of ice present in the ocean, and the effects of climate change on the overall ice coverage worldwide.

14.1 DIFFERENT TYPES OF ICE

The ice observed floating on the surface of the ocean originates from either of two sources. Glacial ice is created by the process of snow accumulation and compression, resulting in the formation of glaciers. These glaciers eventually fracture and release ice into the ocean. Due to the considerable thickness of glaciers, the icebergs that detach from them can attain significant size. Consequently, the towering icebergs found at sea invariably originate from glacial ice sheets. Sea ice is the ice that forms when sea water freezes, and it often does not reach a thickness of a few meters (Figure 14.1.1). At any given moment, around 7% of the ocean is covered with sea ice. This sea ice constitutes about 66% of the Earth's permanent ice cover in terms of area, but only accounts for 0.1% of the total ice volume. This is due to the fact that sea ice is a large expanse of ice that is relatively thin, in contrast to glacial ice caps which are more confined in location yet can be several kilometers thick.

Figure 14.1.1 An iceberg (glacial ice) embedded in a thinner layer of sea ice (NASA / James Yungel [Public domain], via Wikimedia Commons).

The extent of sea ice around Antarctica varies seasonally, ranging from over 21 million km2 during winter to around 1.3 million km2 in the summer. The majority of Antarctic sea ice has a lifespan of only one year. The Arctic experiences relatively modest seasonal variations in ice cover, with a decrease from approximately 14 million km2 during winter to 6.5 million km2 during summer. Approximately 50% of the sea ice in the Arctic endures for a period exceeding one year, transforming into multi-year ice. This disparity occurs because to Antarctica's proximity to water, causing the ice to extend into relatively warmer waters and eventually undergo melting. Due to the surrounding continents, only approximately 10% of the ice from the Arctic Ocean manages to flow into the Atlantic through the passage between Greenland and Spitzbergen. The remaining ice is captured and transforms into multi-year ice or perennial ice, with an average age of approximately 7 years and a thickness ranging from 3 to 5 meters, in contrast to first-year ice which is only 1 to 2 meters thick.

Sea ice formation

Because of the salt content, seawater begins to freeze at about −1.8° C, a lower temperature than for fresh water. Ice formation begins at the surface with the formation of small needle-like ice crystals called **frazil**, which accumulate and make the water appear slushy and cloudy; this stage is referred to as **grease ice** (Figure 14.1.2 A). In calmer water these small crystals can freeze together into a thin surface layer called **nilas**, which can reach a thickness of up to 10 cm (Figure 14.1.2 B).

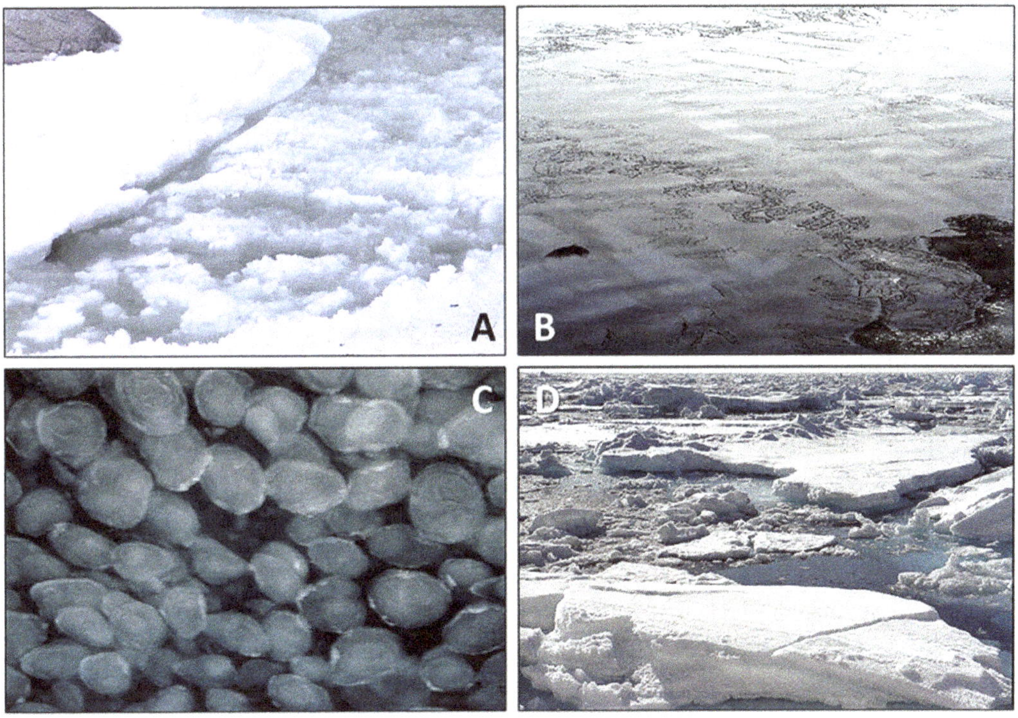

Figure 14.1.2 Stages in sea ice formation. A) grease ice, B) nilas, C) pancake ice, D) ice floes. (A) National Park Service [Public domain], via Wikimedia Commons; B) Brocken Inaglory (Own work) [GFDL (http://www.gnu.org/copyleft/fdl.html) or CC BY-SA 4.0, via Wikimedia Commons; C) PW; D) Jerzy Strzelecki (Own work) [CC BY-SA 3.0], via Wikimedia Commons).

Wave action can break up the nilas into small mats 1-2 m across, which then bump into each other and form rounded shapes with raised edges, called **pancake ice** (Figure 14.1.2 C). If temperatures remain cold the pancake ice freezes together into solid **ice floes**, a hard surface covering the ocean (Figure 14.1.2 D). Ice floes then freeze together into **ice fields**.

Once ice floes form, the water underneath becomes insulated and heat loss to the atmosphere declines, so the water no longer cools and no more ice formation occurs. As a result, young sea ice is usually relatively thin, not more than 3-4 m thick. The ice can get thicker through precipitation; there is not a lot of precipitation at the poles, but due to the low temperatures, whatever does occur tends to accumulate rather than melt. Over time the accumulated ice and snow can add to the overall thickness of the sea ice, but it will still never approach the thickness of glacial ice.

As sea ice crystals form, most of the salt is excluded, so sea ice contains much less salt than seawater, and can be melted for drinking if needed. But about 20% of the salt remains trapped in pockets of water between the ice crystals. As ice forms and salts are excluded into these pockets, the salinity of the remaining water increases and it can become too salty to freeze. These unfrozen pockets of briny water make sea ice a little softer and more slushy than fresh water ice, which is harder and more rigid. Eventually most of this brine leaks out, and the sea ice becomes more solid, but when it is "young ice" it can be more dangerous to walk on than fresh water ice of the same thickness. For example, 7-8 cm of fresh water ice is enough to support the weight of a single person, but you would need at least 15 cm of sea ice to do the same.

The very cold, dense briny water leaks out of the ice and sinks. The brine is "supercooled"; it is cooled below the normal freezing point of seawater, but remains liquid due to the high salt content. When this supercooled brine comes into contact with the surrounding water, it causes the water around it to freeze, creating hollow ice stalactites, or "brinicles" that can be several meters long. Brine continues to flow through the hollow brinicle, and the brinicle grows downwards (see below for an amazing time-lapse video of brinicle formation).

When the large sheets of sea ice are formed, they exist in one of two formations. **Fast ice**, or **land-fast ice**, refers to the large, solid ice sheets that are attached to land. The **pack ice** consists of the smaller, free-floating pieces of sea ice. They may have formed independently, or may have broken off from the fast ice (Figure 14.1.3).

Figure 14.1.3 Fast ice (left) and pack ice (right). (Left: Peterfitzgerald (Own work) [CC BY-SA 3.0], via Wikimedia Commons; Right: Markus Trienke, https://www.flickr.com/photos/mtrienke/34281559366/in/photostream/ [CC BY-SA 2.0]).

The floating pack ice dampens waves and currents, buffering the sea surface from motion. So changes in the distribution of the pack ice can lead to changes in current patterns, and even alter ecosystem structure. But the pack ice is also subject to the currents flowing underneath, and the ice sheets are constantly on the move, breaking up or being pushed together. When pieces of ice converge, they often buckle and crack, or override one another as in convergent lithospheric plate boundaries (section 4.6). These collisions can create tall, jagged pressure ridges, that may extend for several kilometers, and that create hazards for polar explorers navigating the ice (Figure 14.1.4).

Figure 14.1.4 Pressure ridges formed from colliding ice sheets (Michael Studinger [Public domain], via Wikimedia Commons).

In the polar oceans, ice cover is not uniform. There are a number of areas where there is consistently open water, even though the areas surrounding them are covered by ice. These regions of persistent open water are called **polynyas** (Figure 14.1.5). Polynyas may be the result of currents or winds moving the ice, or areas of warmer water that prevent ice formation. In Figure 14.1.5, very strong winds blowing offshore from Antarctica's interior have created a polynya near the edge of the ice sheet.

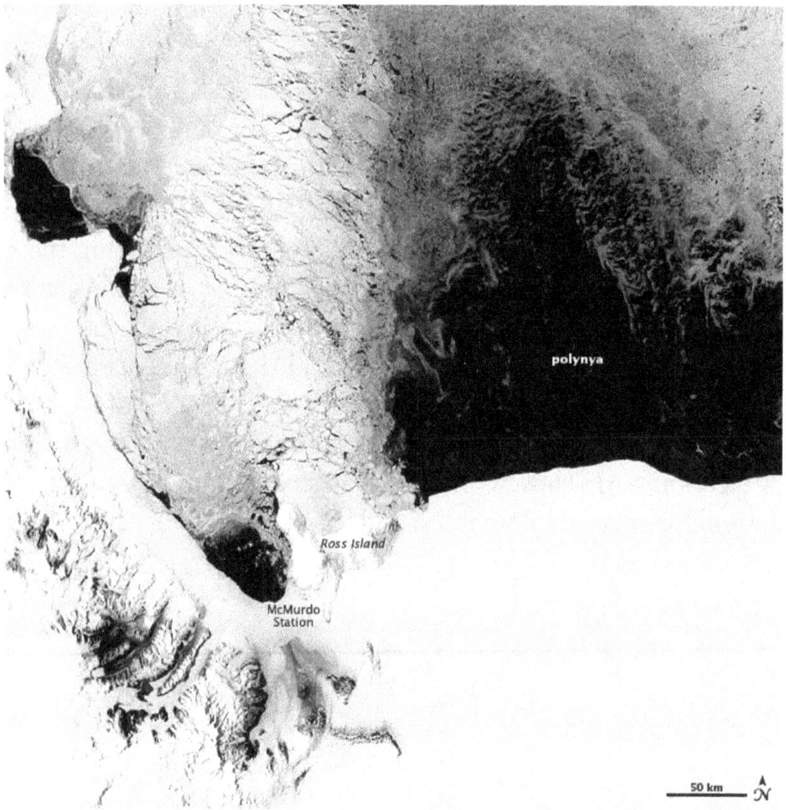

Figure 14.1.5 A polynya near McMurdo Station in Antarctica (NASA's Earth Observatory [CC BY 2.0], via Wikimedia Commons).

14.2 ICEBERGS

Icebergs are among the most magnificent ice formations found in the ocean. It is important to note that the largest icebergs are not formed from sea ice. Instead, they are chunks of glaciers that have broken off, or calved, from the glacier tongue. Therefore, these icebergs are made up of fresh water ice.

Glacial ice is created through the gradual accumulation of layers of snow. The weight of the accumulated layers causes the snow to be compressed into a granular substance called firn, which has a greater density and less air compared to normal snow. As the snow continues to accumulate, the pressure on it increases, causing the firn to be squeezed and transformed into denser glacial ice (Figure 14.2.1). The predominant color of most icebergs is white due to the presence of numerous air bubbles within the ice, which induce the scattering of all wavelengths of white light. However, icebergs made of older ice or densely packed ice from the inner parts of a glacier can have a distinct and intense blue color (as shown in Figure 14.2.1). This ice has a lower amount of air and is made up of bigger, more compact ice crystals that soak up the longer (red) wavelengths of light and allow the shorter, blue wavelengths to pass through and scatter. As the distance that light travels through the ice increases, the absorption of longer wavelengths also increases, resulting in a bluer appearance of the ice.

Figure 14.2.1 Left: Changes in ice crystal structure and density as they are compacted from snow into glacial ice (Steven Earle, "Physical Geology"). Right: A blue iceberg (By Doug Knuth [CC BY-SA 2.0], via Wikimedia Commons).

Icebergs come in a wide range of sizes, from a few meters across to hundreds of kilometers long. Table 14.2.1 shows the size categories used in the North Atlantic (in the Antarctic a different scale is used, as the icebergs tend to be larger there).

Table 14.2.1 Iceberg size classifications for the North Atlantic

Size Category	Height (m)	Length (m)
Growler	< 1	< 5
Bergy Bit	1-5	5-15
Small	5-15	15-60
Medium	16-45	61-120
Large	46-75	121-200
Very Large	> 75	> 200

Icebergs can also be categorized according to their forms (Figure 14.2.2). The main differentiation lies between tabular and non-tabular icebergs. Tabular icebergs possess vertical sides and a level surface, with a length that exceeds five times their height. Non-tabular icebergs refer to icebergs that do not have a tabular shape. These icebergs can be further classified into several groups. Domed icebergs exhibit a rounded apex, pinnacled icebergs include towering spires, and wedge icebergs possess a steep facade adjacent to a gently sloping side. The drydock icebergs include a water-filled passage that is theoretically spacious enough to accommodate the passage of boats, hence the name. Blocky icebergs possess a level surface and steep slopes, however their proportion of length to height is not as significant as that of a tabular iceberg. The visible portion of an iceberg, regardless of its shape, accounts for only 10% of its total mass, while the other ice remains beneath.

Figure 14.2.2 Icebergs can be classified based on their shapes. Clockwise from top left: tabular, domed, pinnacled, blocky, drydock, and wedge icebergs. (Tabuar: Michael Clarke Stuff [CC BY-SA 2.0 via Wikimedia Commons]; domed: AWeith [CC BY-SA 4.0] via Wikimedia Commons; pinnacled: Povl Abrahamsen, https://www.flickr.com/photos/povl/13339227703 [CC-BY-NC 2.0]; wedge: https://pixabay.com/en/iceberg-st-john-s-newfoundland-2008688/ [CC0 Public domain]; drydock: Brocken Inaglory [CC-BY-SA-3.0] via Wikimedia Commons; blocky: Jerzy Strzelecki [CC-BY 3.0] via Wikimedia Commons).

How much of an iceberg is visible above the surface?

Archimedes' Principle states that the upwards buoyant force of an object in water is equal to the weight of the water displaced by the object. We know that the density of an iceberg (ρ_i) is around 0.917 g/cm^3. The weight (w_i) of the iceberg is equal to its mass (m_i) x the acceleration due to gravity (g), which is 9.8 m/s^2. Since density = mass/volume (V), the weight of the iceberg is:

$$w_i = \rho_i V_i g$$

Since the iceberg is floating at equilibrium, the weight of the iceberg (w_i) is equal to the weight of the water it displaces (w_w). The weight of water displaced is therefore:

$$w_w = \rho_w V_w g$$

where the density of seawater (ρ_w) is about 1.024 g/cm^3. Now we have:

$$W_i = W_w$$

$$\rho_i V_i g = \rho_w V_w g$$

$$\text{so } V_w = (\rho_i/\rho_w)V_i = (0.917/1.024)V_i = 0.89\ V_i.$$

In other words, the volume of water displaced (V_w) is equal to about 89% of the volume of the iceberg (V_i). This means that 89% of the iceberg is submerged, leaving around 11% of the ice exposed above the surface.

Icebergs in the Arctic tend to be smaller and more non-tabular than Antarctic icebergs, because most of them formed as irregular chunks of glaciers that calved off into the ocean, primarily around Greenland. It is estimated that the glaciers around Greenland and the Canadian Arctic calve off 300 billion cubic meters of icebergs each year. Because Antarctica has much larger ice sheets, the icebergs produced in the Southern Hemisphere are usually larger than those in the Arctic, and they tend to be more tabular in shape, as entire sections of the ice sheet break off at once. The largest iceberg ever recorded, the B-15 iceberg, broke off of the Antarctic ice sheet in 2000. It measured 295 km long and 37 km wide, giving it an area of about 11,000 square kilometers; about the size of Connecticut. Over the next few years B-15 drifted along the Antarctic coast before being broken up into smaller pieces (Figure 14.2.3).

Figure 14.2.3 Left: Iceberg B-15 soon after breaking off of the Antarctic ice sheet in 2000 (NASA/GSFC/Robert Bindschadler [Public domain], via Wikimedia Commons). Top right: Aerial view of a section of B-15A, a piece of the B-15 iceberg in January 2002 (NSF/Josh Landis, [Public domain], via Wikimedia Commons). Bottom right: Map showing the path of the B-15A iceberg over several years (Luca Pietranera [Public domain], via Wikimedia Commons).

All of these icebergs floating through the polar oceans are potentially hazardous to ship traffic, as we all know from the tragic story of the Titanic. In response to that disaster, the North Atlantic maritime nations established the International Ice Observation and Ice Patrol Service in 1914, to monitor icebergs threatening the shipping lanes. Interestingly, the original charge of the patrol also included searching for and destroying derelict ship hulls that were still floating around, as these were also a potential hazard. While the patrols were successful at destroying these abandoned ship hulls, they achieved far less success at attempting to destroy the icebergs. Eventually they gave up on iceberg destruction and focused their efforts on monitoring iceberg movements.

Today, the service still operates as the International Ice Patrol, operated by the U.S. Coast Guard. Twice each day the patrol issues alerts with the position and potential tracks of existing icebergs, and the extent of the ice cover, reporting on the roughly 600 icebergs per year that intrude below 48° N latitude, the northern limit of the major shipping lines. No such iceberg patrol exists around Antarctica, as there is much less shipping traffic at those latitudes.

14.3 ICE AND CLIMATE CHANGE

In section 8.5 we examined the causes of climate change, and the potential impacts on Earth's ecosystems. One obvious consequence of increasing global temperatures is a decline in ice cover as the warmer climate melts the ice caps. So are we seeing a decrease in ice cover, and if so, what are some consequences of this decline?

To address these questions we must once again begin by separating sea ice and glacial ice, as each type of ice is responding to climate change in its own way, with different consequences. For example, melting polar ice is often discussed in terms of its effect on raising sea levels (section 13.7). Sea level is not really impacted by the melting of sea ice, as the formation and melting of sea ice does not change the volume of water already in the ocean. The melting of glacial ice, on the other hand, takes water that was locked up in glaciers on land and returns it to the ocean. This does increase the volume of water in the oceans, and results in rising sea levels. So we will first examine the trends in sea ice cover, and then look at the situation with glacial ice.

While fluctuations in sea ice cover do not dramatically impact sea level, they can be used as indicators of global warming. Satellite data show that since 1979 the average Arctic sea ice extent declined by about 3% per decade, with the rate of decline increasing in recent years (Figure 14.3.1). The year 2017 set a record for the lowest maximum Arctic sea ice extent in the 38-year history of the satellite data, the third year in a row that the record was broken. In addition, the thickness of the Arctic sea ice has also declined, from a mean thickness of 3.64 meters in

1980 to 1.89 meters in 2008. So Arctic sea ice cover is clearly in decline as global temperatures increase.

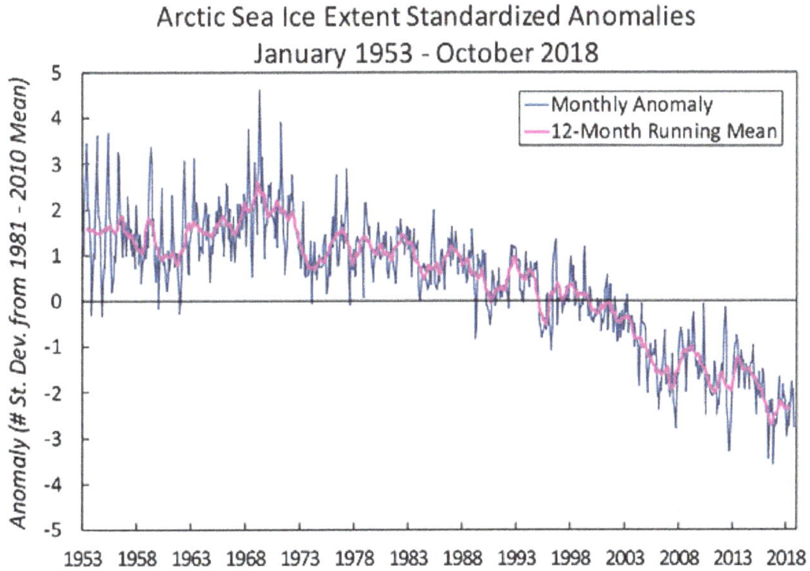

Figure 14.3.1 Mean sea ice anomalies from 1953-2018. Positive values indicate ice cover greater than the mean from 1981-2010, negative values show ice cover below that mean (Image by Walt Meier and Julienne Stroeve, National Snow and Ice Data Center).

The status of Antarctic sea ice is ambiguous. The extent of sea ice in the Antarctic is seeing a minor increase, with a growth rate of approximately 1.1% per decade. However, it is important to note that this upward trend is only about one-third of the magnitude of the decreasing trend observed in the Arctic (Figure 14.3.2). There are other possibilities on the cause of this increase, such as the influence of winds, currents, and La Niña conditions (section 9.6) in the Pacific Ocean. As an illustration, the robust Antarctic Circumpolar Current acts as a barrier that hinders the flow of warmer water towards the ice edge. Additionally, the prevailing winds in the area maintain low temperatures, which facilitate the production of ice. Another plausible explanation is that while certain regions of the Antarctic experience rising temperatures, there is a corresponding increase in precipitation and subsequent glacial ice melting. As a result, a stratum of less dense, more pure water forms on the surface and freezes more readily than water with higher salt content. In the Arctic, rivers have consistently contributed a substantial amount of freshwater, resulting in the presence of a low-density layer. Regardless of the causes for the growing sea ice in the Antarctic, it is important to note that the increase is still smaller than the decrease observed in the Arctic. Moreover, the general trend for world sea ice indicates a loss of approximately 35,000 km2 each year since 1979.

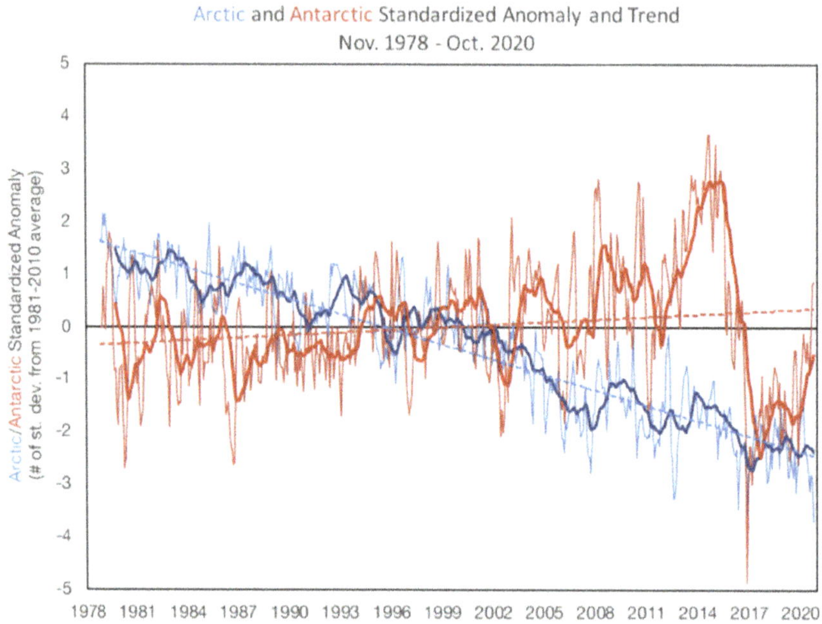

Figure 14.3.2 Arctic and Antarctic Sea Ice Extent Anomalies, 1978-2020: Arctic sea ice extent underwent a strong decline from 1978 to 2020, but Antarctic sea ice underwent a slight increase, although some regions of the Antarctic experienced strong declining trends in sea ice extent (National Snow and Ice Data Center).

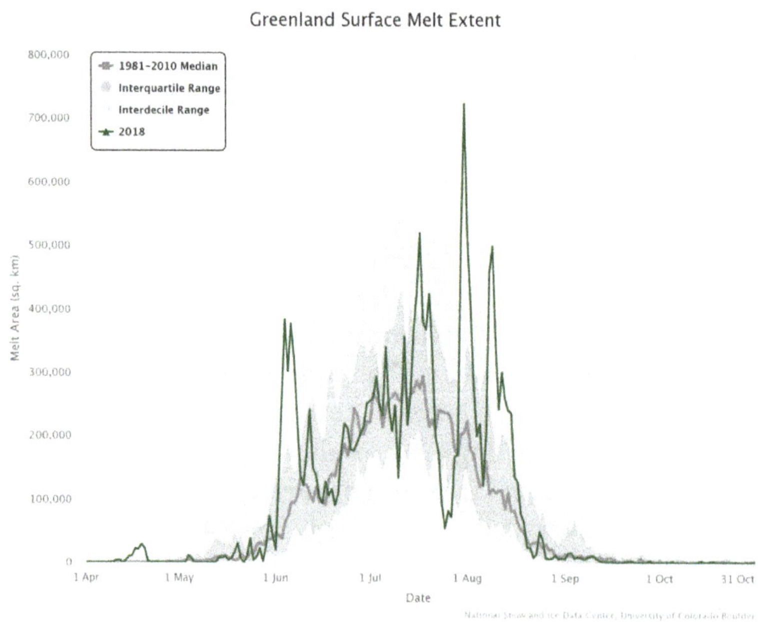

Figure 14.3.3 2018 ice melt in Greenland. For much of the year, ice melt was significantly higher than the 1981-2010 average (National Snow and Ice Data Center).

Regarding glacial ice, the Arctic is once again displaying noticeable indications of substantial ice depletion, particularly in Greenland. Between 1979 and 2006, there was a 30% rise in summer melting on the Greenland ice sheet. The ice melt seen in 2016 ranked as the tenth largest on record, as shown in Figure 14.3.3. Once again, the patterns are less distinct in Antarctica. The West Antarctic ice sheet and the Antarctic Peninsula are experiencing both warming and ice loss. The Antarctic Peninsula has warmed by 2.5 degrees Celsius since 1950. Nevertheless, East Antarctica does not exhibit a discernible pattern of increasing temperatures, and certain regions within this area are actually experiencing a buildup of ice. In general, the Antarctic ice is presently pretty steady, especially in comparison to Greenland. The melting of the polar ice sheets would significantly affect the sea level. The anticipated impact of Greenland ice melting on global sea level is approximately 7 meters, but the entire melting of the Antarctic ice sheet would lead to a rise of nearly 60 meters.

GLOSSARY

abyssal plain: the flat seafloor of the deep ocean, typically beyond the limits of the continental slopes

abyssal zone: the region of the seafloor between 4000-6000 m

abyssopelagic zone: the deeper parts of the open ocean, between 4000 and 6000 m; also known as the abyssalpelagic zone

accretion (planetary): the process by which solid celestial bodies are added to existing bodies during collisions

active continental margin: where the boundary between the continent and the ocean is also a tectonic plate boundary

aeolian: processes related to transportation and deposition of sediments by wind

albedo: the reflectivity of a surface of a planet (expressed as the percentage of light that reflects from the surface)

amphidromic circulation: gyre-sized tidal patterns where a tidal crest rotates around an ocean basin

amphidromic point: the center point around which amphidromic circulation rotates; there is near-zero tidal range at the amphidromic point (also called a tidal node or amphidromic node)

anoxia: conditions of zero or extremely low dissolved oxygen, usually below 0.5 mg/L

Antarctic Bottom Water: water at abyssal depths in the ocean that forms from the sinking of dense cold water adjacent to Antarctica

Anthropogenic: resulting from the influence of humans

aphotic zone: depths beyond 1000 m where there is no light penetration

arch: a rock weathering remnant in the form of an arch (typically along a coast and resulting from wave erosion)

asthenosphere: the part of the mantle, from about 100 to 200 km below surface, within which the mantle material is close to its melting point, and therefore relatively weak

atmospheric wave: a wave formed in the atmosphere at the boundary between air masses of different densities

atoll: a ring-shaped carbonate (or coral) reef or series of islands

backshore: the region of the beach above the high tide line, which is only submerged under unusually high wave conditions

backwash: the wash of wave water down the slope of a beach

bar-built estuary: an estuary created when a sand bar or barrier island cuts off the estuary from mixing completely with seawater

barrier island: a long, thin island parallel to the shore, created through the deposition of sand

barrier reef: a reef that forms a barrier to waves along a coast; it is separated from land by a lagoon

barycentre: the center of mass in the Earth-moon system around which they rotate

basalt: a volcanic rock that makes up much of the oceanic crust

bathyal zone: the region of the seafloor from the shelf break to 4000 m

bathymetry: pertains to measuring the depths of the ocean

bathypelagic zone: the moderately deep parts of the open ocean, between 1000 and 4000 m

baymouth bar: a spit that extends across the mouth of a bay

beach face: the area of a beach between the high and low tide lines

Beaufort scale: a 0-12 scale describing the wind conditions at sea, often reflected in wave heights

benthic: refers to the environment of the seafloor

benthos: refers to the community of organisms living on or in the ocean floor

berm: a flat area of a beach in the backshore area (above the high tide level)

big-bang theory: the theory that the universe started with a giant expansion approximately 13.77 billion years ago

biogenous sediment: sediment created from the remains of organisms

blocky iceberg: iceberg with a flat top and steep sides, but their length to height ratio is not as great as it is for a tabular iceberg

blowhole: a hole in the ceiling of an arch or sea cave through which water is ejected when waves approach

body wave: a seismic wave that travels through rock (e.g., a P-wave or an S-wave)

boulder: a sediment with a grain diameter of at least 256 mm

boundary currents: ocean currents whose properties are influenced by the presence of a coastline

Boyle's Law: the volume of a gas is inversely proportional to the pressure

brackish: seawater of low salinity; part fresh water, part seawater

breaker: an unstable wave that has collapsed

breakwater: a structure built offshore in order to deflect the energy of waves

buffer: a solution that moderates changes in pH when acids or alkalis are added to it

calcareous sediment: sediments composed of calcium carbonate, often from the shells of marine organisms

calving: when ice breaks off of the front of a glacier and collapses into the water

capillary waves: small ripples that form on the water surface under light winds; their restoring force is surface tension

carbonate compensation depth: the depth in the ocean (typically around 4000 m) below which carbonate minerals are soluble

Carboniferous: a geologic period that spans 60 million years from the end of the Devonian Period 358.9 million years ago, to the beginning of the Permian Period, 298.9 Mya

celerity: the speed of a wave

chemosynthesis: the creation of organic compounds using the energy from inorganic chemical reactions

chip log: a device for determining a ship's speed at sea, by measuring the rate at which a line is unspooled when cast overboard

clay: sediment particle that is less than 1/256 mm in diameter

climate feedback: a process by which the physical effects of a climate forcing can have other effects (either negative or positive) on the climate

climate forcing: a mechanism, such as a change in greenhouse gas levels, that forces the climate to change

coastal plain estuary: an estuary formed when sea level rises and submerges a river valley (also known as a drowned river valley estuary)

coastal straightening: the tendency for an irregular coast to be straightened over time by coastal erosion processes

cobble: sediment particle that is between 64 and 256 mm in diameter

coccolithophore: photosynthetic algae that makes its test (shell) out of calcium carbonate

compensation depth: the depth where the rate of photosynthesis equals the rate of respiration

conduction: the transfer of heat through direct contact

conservative ions: ions whose proportions are the same regardless of overall salinity; the major ions in seawater

constructive interference: where the interaction of multiple waves creates waves larger than any of the component waves

continental crust: the Earth's crust underlying the continents (as opposed to ocean crust)

continental drift: the idea that the continents have moved over the surface of the Earth over geological time

continental margin: the region of transition from the land to the deep sea floor, i.e. between continental and oceanic crust

continental rise: the area at the bottom of the continental slope, where it transitions to the abyssal sea floor

continental shelf: the shallow (typically less than 200 m) and flat sub-marine extension of a continent

continental slope: the steeper part of a continental margin, that slopes down from a continental shelf towards the abyssal plain

convection cell: a rotating region in a fluid in which upward motion of warmer, low density fluid in the center is balanced by downward motion of cooler, denser fluid at the periphery

convergent boundary: a plate boundary at which the two plates are moving towards each other

core: the metallic interior part of the Earth, extending from a depth of 2900 km to the center

core-mantle boundary: the boundary, at 2900 km depth, between the mantle and the core

Coriolis Effect: the tendency for the path of moving bodies (e.g., ocean currents) to be deflected on the surface of the Earth, to the right in the Northern Hemisphere and to the left in the Southern Hemisphere

cosmogenous sediment: sediment derived from extraterrestrial sources

crest: the highest point on a wave

Cretaceous: a geologic period that spans 79 million years from the end of the Jurassic Period 145 million years ago to the beginning of the Paleogene Period 66 mya

crust: the uppermost layer of the Earth, ranging in thickness from about 5 km (in the oceans) to over 50 km (on the continents)

deep water wave: a wave above a water depth greater than half of its wavelength

delta: large, often triangular accumulation of sediment near the mouth of a river

denitrification: where nitrate is converted to molecular nitrogen through a series of intermediate nitrogen oxide products

density: mass per unit volume of a substance (e.g., g/cubic cm)

destructive interference: where the interaction of multiple waves creates waves smaller than any of the component waves

diatom: photosynthetic algae that make their tests (shells) from silica

diatomaceous earth: powdery sediment composed of silica diatom tests

differentiation: the un-mixing of a magma, typically by the physical separation of minerals that crystallize early and settle towards the bottom

dinoflagellate: photosynthetic algae characterized by the presence of flagella and a cellulose test (shell)

discoaster: an extinct form of single-celled algae that produced calcareous tests that can still be found in some marine sediments

diurnal tide: a tidal cycle with only one high and one low tide per day

divergent boundary: a plate boundary at which the two plates are moving away from each other

doldrums: areas of low pressure and weak winds along the equator

domed iceberg: iceberg with a rounded top

downwelling: process by which surface water is forced downwards

drowned river valley estuary: an estuary formed when sea level rises and submerges a river valley (also known as a coastal plain estuary)

drydock iceberg: iceberg with a water-covered channel running through it

dysphotic zone: depths of the water column where there is some light penetration, but not enough to support photosynthesis; corresponds to the mesopelagic zone, 200-1000 m. Also known as the twilight zone

ebb current: current created by an outgoing tide

eccentricity: in the context of Milankovitch Cycles, the degree to which the Sun is offset from the geometric center of the Earth's orbit

eddy: a rotating water mass

Ekman spiral: where each layer of water is deflected relative to the layer above it, forming a spiral that extends down to about 100 m

Ekman transport: bulk transport of water due to the Ekman spiral; the net movement Ekman transport is 90 degrees relative to the wind direction

El Niño: a periodic climatic situation in which warm water extends all or most of the way to the eastern edge of the equatorial Pacific

El Niño-Southern Oscillation (ENSO): the fluctuating atmospheric conditions that lead to the localized ocean warming of El Niño

electron: a sub-atomic particle of essentially no mass and a single negative charge

Eocene: a geological epoch lasting from 56 to 33.9 million years ago

epicentre: the location on the surface vertically above the location (i.e., "hypocenter" or "focus") where an earthquake takes place

epipelagic zone: the upper layer of water (0 to 200 m) in areas of the open ocean

estuary: a partially enclosed body of water where seawater is diluted by freshwater input

euphotic zone: the upper regions of the ocean where there is enough light to support photosynthesis; approximately 0-200 m; also called the photic zone

eustatic sea level change: sea level change related to a change in the volume of the oceans, typically because of an increase or decrease in the amount of glacial ice on land

evaporates: hydrogenous sediments that form when seawater evaporates

fast ice: ice sheets that are attached to land

fault: a boundary in rock or sediment along which displacement has taken place

fecal express: small particles reach the seafloor much faster when incorporated into large fecal pellets than if they sank on their own

feedback: a process by which the physical effects of a climate forcing can have other effects (either negative or positive) on the climate

Ferrel Cell: the atmospheric convection cells between 30 and 60 degrees latitude

fetch: the distance over which wind blows to form waves

firn: the granular transitional state between snow and ice within a glacier

fjord: a deep, U-shaped estuary that was carved out by advancing glaciers

flood current: current created by an incoming tide

flushing time: the time it would take for all of the fresh water in an estuary to be replaced by runoff of new water

focus (earthquake): the actual point below surface at which an earthquake takes place (equivalent to hypocenter)

foraminifera: a single-celled protist with a shell that is typically made of calcium carbonate

foreshore: the part of a beach between the high tide and low tide lines

frazil: small, needle-like crystals in the first stages of sea ice formation

frequency: the number of waves that pass a point in a given amount of time

fringing reef: a reef adjacent to a shoreline where there is either a very narrow back reef area or none at all (in which case the reef is effectively attached to the shore)

frost line: in the context of planetary systems the boundary beyond which volatile components (e.g., water, carbon dioxide, methane, ammonia etc.) are frozen

Ga: (gigaannus) billions of years before the present

galaxy: a gravitationally-bound system of stars and interstellar matter

gas giant: a large planet composed mostly of hydrogen and helium (e.g. Jupiter)

geostrophic flow: circular currents created from the balance between gravity- and Ekman-driven flow

giant impact hypothesis: the theory that the Moon formed when a Mars-sized planet (Theia) collided with the Earth at 4.5 billion years ago

glacial groove: scratches and grooves carved into bedrock from rocks carried by moving glaciers

glacial ice: ice formed from the accumulation and compression of snow into glaciers

glacial period: a period of Earth's history during which glacial ice was present over a sufficient extent to have left recognizable evidence

glacier: a long lasting (centuries or more) body of ice on land that moves under its own weight

granite: an igneous (formed from cooling magma) rock that comprises much of the continental crust

granule: a sedimentary particle ranging in size from 2 to 4 mm in diameter

grease ice: an accumulation of frazil to create a slushy consistency in sea ice formation

greenhouse effect: in the context of climate, the ability of an atmosphere to absorb infrared radiation due to the presence of greenhouse gases

greenhouse gas: a gaseous molecule with 3 or more atoms that is able to absorb infrared radiation

groin (groyne): a man-made structure extending from the shore built to deflect the energy of waves

groin field: a series of groins along a beach

gross primary production: the total amount of organic material created by primary producers

groundwater: water that lies beneath the surface of the ground

Gulf Stream: the major surface current flowing northwards along the Atlantic coast of the U.S. and Canada

guyot: a flat-topped seamount (also called a tablemount)

gyre: a large circular ocean surface current

hadal zone: the region of the seafloor below 6000 m

Hadley Cell: the atmospheric convection cells between the equator and 30 degrees latitude

hadopelagic (hadalpelagic) zone: region of the open ocean with water depths greater than 6000 m

halite: NaCl, a mineral also known as table salt

halocline: where there is a dramatic change in salinity over a small change in depth

hard stabilization: the building of physical structures to prevent the erosion of beaches and shorelines

harmful algal bloom (HAB): when phytoplankton appear in very high concentrations with potentially hazardous consequences such as mass die-offs or toxicity

headland: a point of land extending out to sea

heat budget: the balance between the amount of heat entering and leaving the Earth

heat capacity: the amount of heat needed to change a substance's temperature by one degree

Henry's Law: as the pressure increases, a fluid will contain more dissolved gas

high pressure: in atmospheric terms, a region of descending air, increasing the atmospheric pressure. Winds blow away from high pressure zones

highly stratified estuary: a deep estuary with some mixing at the surface, but little mixing at depth

homolosine projection: a map projection where area is retained, but there are interruptions to the continents or oceans

horse latitudes: areas of high pressure and weak winds around 30 degrees latitude in both hemispheres

hot spot: the surface area of volcanism and high heat flow above a mantle plume

hydrogen bond: a weak bond between two molecules due to the electrostatic attraction of a proton in one molecule to the negative polar end of the other molecule

hydrogenous sediment: sediments formed from the precipitation of dissolved substances

hydrological cycle: the cycling of water through the ocean, atmosphere, lakes, organisms, and other reservoirs

hydrothermal vent: area of the seafloor where superheated water seeps out of the crust

hypocentre: the actual point below surface at which an earthquake takes place (equivalent to focus)

hypoxia: a condition with low dissolved oxygen, usually defined as oxygen levels below 2 mg/L

ice field: an area covered by ice floes

ice floe: a relatively large piece of floating sea ice

ice giant: a planet that is comprised mainly of gases heavier than hydrogen and helium, including oxygen, carbon, nitrogen, and sulfur (e.g., Uranus and Neptune)

iceberg: a large, floating piece of glacial ice

inner core: the solid metal mass at the center of the Earth, extending 1200 km from the center

insolation: a measure of the intensity of solar energy at a specific location or time (expressed in W/square m)

Intergovernmental Panel on Climate Change (IPCC): an international body established in 1988 by the UN's World Meteorological Organization and the UN Environment Program to prepare periodic reports on the status of global climate change and its mitigation

intermediate wave: a wave in a water depth between 1/2 and 1/20 its wavelength

internal wave: waves that form below the surface at the interface between water masses of different densities

intertidal zone: the region of a coast between the high and low tide lines. Also called the littoral zone

Intertropical Convergence Zone: the area near the equator where the northeast and southeast trade winds converge; known for weak winds, it is also called the doldrums

Ion: an atom or molecule that has either gained or lost electrons and has thus become charged

island arc: long chains of volcanic islands found along convergent tectonic plate boundaries

isostasy: the equilibrium position reached by a block of crust floating on the underlying fluid mantle

isostatic sea level change: the effect on relative sea level of a vertical movement of the crust resulting from a change in the mass of the crust (e.g., from losing or gaining ice)

isotope: forms of the same element that contain equal numbers of protons but different numbers of neutrons in their nuclei

jetty: a long structure built to protect a harbor from filling with sand due to longshore transport

Jovian planet: a gas giant

Ka: (kiloannus) thousands of years before the present

kinetic energy: the energy that an object possesses due to its motion

knot: one knot (kt) = 1 nautical mile per hour = 1.15 mph = 1.85 kph

La Niña: a periodic climatic situation in which colder than normal water extends throughout the equatorial Pacific

land breeze: winds blowing from land towards the ocean

Langmuir circulation: corkscrew circulation patterns formed parallel to strong winds; they only extend a few meters below the surface

latent heat of fusion: the heat required to change a substance from solid to liquid; 80 cal/g in the case of ice melting to water

latent heat of vaporization: the heat required change a substance from liquid to gas; 540 cal/g to turn water into vapor

latitude: the distance north or south of the equator, measured as an angle from the equator

Laurentide Ice Sheet: the continental glacier that extended across central eastern North America during the Pleistocene, covering most of Canada and a significant part of the United States

Lithification: the conversion of unconsolidated sediments into rock by compaction and cementation

lithogenous sediment: sediment derived from preexisting rock

lithosphere: the rigid outer part of the Earth, including the crust and the mantle down to a depth of about 100 km

littoral drift: the movement of sediment along a shoreline resulting from a longshore current and also from the swash and backwash on a beach face (another name for longshore transport)

littoral zone: the region of a coast between the high and low tide lines. Also called the intertidal zone

longitude: measurement of distance east or west of the prime meridian, expressed as an angle

longshore bar: an offshore deposit of sand parallel to the shoreline

longshore current: the movement of water parallel to a shoreline produced by the approach of waves at an angle to the shore

longshore transport: the movement of sediment along a shoreline resulting from a longshore current and also from the swash and backwash on a beach face. Also known as littoral drift

low pressure: in terms of the atmosphere, a region of rising air, lowering the atmospheric pressure. Winds blow towards low pressure regions, which are often characterized by precipitation from rising, cooling, condensing air (8.3)

low tide terrace: another name for the beach face

lysocline: the depths where the rate of calcium carbonate dissolution increases dramatically over surface waters

Ma: (Megaannus) millions of years before the present

magma: molten rock typically dominated by silica

magnetic dip: the angle of the magnetic field within a rock, relative to the horizontal; may be used to infer the latitude where the rock was first formed

magnitude: a measure of the amount of energy released by an earthquake

major ions: the six ions that comprise over 99% of the ions in the ocean (chloride, sodium, sulfate, magnesium, calcium, potassium)

manganese nodule: spherical accumulations of manganese and other metals that form slowly through precipitation on the seafloor

mantle: the middle layer of the Earth, dominated by iron and magnesium rich silicate minerals and extending for about 2900 km from the base of the crust to the top of the core

mantle convection: movements in the mantle from rising and sinking mantle material as it heats and cools

mantle plume: a plume of hot rock (not magma) that rises through the mantle (either from the base or from part way up) and reaches the surface where it spreads out and also leads to hot-spot volcanism

meander: the sinuous path taken by a current, such as the Gulf Stream

Mercator projection: a map projection where latitude and longitude are both represented as straight, parallel lines intersecting at right angles

mesopelagic zone: the upper middle zone of the open ocean extending from 200 to 1000 m depth

Mesozoic: the geological era from about 252 to 66 million years ago

meteoroid: a fragment of either stony or metallic debris in space

methane hydrate: a combination of water ice and methane in which the methane is trapped inside "cages" in the ice

mid-ocean ridge: an underwater mountain system along divergent plate boundaries, formed by plate tectonics

Milankovitch cycles: millennial-scale variations in the orbital and rotational parameters of the Earth that have subtle effects on the Earth's climate

mixed interference: where the interaction of multiple waves creates both constructive and destructive interference and an irregular surface pattern

mixed layer: the topmost layer of the ocean, where winds, waves, and currents mix the water so that conditions are relatively constant; approximately the top 100 m

mixed semidiurnal tide: a tidal cycle with two high and two low tides per day, each of different heights

Mohorovičić discontinuity (Moho): the boundary between the crust and the mantle

nautical mile: a distance equal to one minute of latitude; equivalent to 1.15 land miles or 1.85 km

neap tide: the period of minimum tidal range when the Earth is perpendicular to the sun and moon

nearshore: the part of a beach from the low tide line to the depth where wave action is no longer influenced by the bottom, i.e. to where the depth exceeds the wave base

nebula: a cloud of interstellar dust and gases

negative feedback: a process that results in a decrease in that process (in the context of climate change it is a process that reduces the change in climate, such as the enhanced growth of vegetation in response to an increase in atmospheric carbon dioxide)

neritic: the marine pelagic province from the low tide line to the shelf break (1.3)

net production: total primary production minus the organic compounds used up by respiration by the producers

new production: primary production supported by nutrients brought in from outside of the local ecosystem

nilas: a thin surface sheet of sea ice

nitrification: the biological oxidation of ammonia or ammonium to nitrite followed by the conversion of the nitrite to nitrate

nitrogen fixation: the process of converting atmospheric nitrogen gas into nitrogenous compounds like ammonia

non-conservative ions: ions in seawater whose proportions fluctuate with changes in salinity

non-tabular iceberg: an iceberg with any shape other than tabular

North Atlantic Deep Water: deep Atlantic Ocean water that has descended in the far north of the basin in the area between Scandinavia and Greenland

nutrient: in the context of primary production, substances required by photosynthetic organisms to undergo growth and reproduction

nutrient-like element: elements that have a similar vertical profile to nutrients; low amounts at the surface, increased abundance at depth

obliquity: in the context of Milankovitch Cycles, the angle of the tilt of the Earth's rotational axis with respect to the plane of its orbit around the Sun

ocean acidification: where the overall pH of the ocean declines, likely due to an increased amount of carbon dioxide in the ocean

oceanic: the marine pelagic province representing the open ocean regions, i.e. beyond the neritic zone

oceanic crust: the Earth's crust underlying the oceans (as opposed to continental crust)

offshore: the beach zone beyond the nearshore region

oolite: a small (approximately 1 mm) sphere of calcite formed in areas of tropical shallow marine water

ooze: a sediment composed of >30% biogenous material

outer core: the layer of the inner Earth extending 2300 km from the top of the inner core to the bottom of the mantle, composed of fluid metal alloys

outgassing: where dissolved substances in magmas are released as gases when the pressure is reduced

overturning: the vertical cycling within a body of water, where denser water sinks and less dense water floats to the surface

oxygen minimum layer: region of ocean depths where dissolved oxygen is at its lowest level; usually around 1000 m for the open ocean

p-wave: a seismic body wave that is characterized by deformation of the rock in the same direction that the wave is propagating (compressional vibration)

pack ice: free-floating ice floes

Paleocene: a geological epoch that lasted from about 66 to 56 million years ago

paleomagnetic: past variations in the intensity and polarity of the Earth's magnetic field

Paleozoic: the geologic era lasting from 541 to 252 million years ago

pancake ice: small, rounded, thin pieces of sea ice that will freeze together to form an ice floe

Pangaea: the supercontinent that existed between approximately 300 and 180 Ma; it contained all of the modern continents combined into a single land mass

partially mixed estuary: where salinity increases from the head to the mouth, but there is also a slight increase in salinity with depth at any point; also called a slightly stratified estuary

passive continental margin: a boundary between a continent and an ocean at which there is no tectonic activity (e.g., the eastern edge of North America)

pebble: a sedimentary particle ranging in size from 2 to 64 mm (includes granule)

pelagic: relating to the open ocean

period: the time it takes for a complete wave to pass a given point

Permian: a geologic period which spans 47 million years from the end of the Carboniferous Period 298.9 million years ago, to the beginning of the Triassic period 251.902 Mya

Phanerozoic: the eon on the geological time scale covering time from the beginning of the Cambrian period 541 million years ago to the present, and comprising the Paleozoic, Mesozoic, and Cenozoic eras

phase change: the change of state between a solid, liquid, or gas

photic zone: the upper regions of the ocean where there is enough light to support photosynthesis; approximately 0-200 m; also called the euphotic zone

photosynthesis: the production of organic compounds from carbon dioxide and water, using sunlight as an energy source

physiographic projection: map projection presenting bathymetry or altitude data as a 3D relief map

phytoplankton: drifting, usually single-celled algae that undergo photosynthesis

picoplankton: planktonic bacteria

pinnacled iceberg: an iceberg with one or more tall spires

planisphere projection: map projection that keeps latitude horizontal, but shows some convergence of longitude

plankton: an organism that cannot swim effectively, so it drifts with the currents

plate: a region of the lithosphere that is considered to be moving across the surface of the Earth as a single unit

plate tectonics: the concept that the Earth's crust and upper mantle (lithosphere) is divided into a number of plates that move independently on the surface and interact with each other at their boundaries

plunging breaker: a breaking wave on moderately-steep beaches that curls over on itself as it breaks

Polar Cell: the atmospheric convection cells between 60 degrees latitude and the pole

polar easterlies: the dominant wind bands between the poles and 60 degrees latitude

polar front: the boundary between the polar cell and the Ferrel cell around the 60 latitude in each hemisphere

polar molecule: a molecule where the electrons are not distributed equally, leading to a charge imbalance across the molecule; portions of the molecule are slightly positive while other portions are slightly negative

polar wandering path: a path of varying magnetic pole positions defined by paleomagnetic data (in fact it is now understood that the continents have wandered, not the poles, so a more appropriate terms is "apparent polar wandering path")

polyna: an area of persistent open water in areas otherwise covered with ice (14.1)

positive feedback: a process that results in an increase in that process (in the context of climate change it is a process that enhances the change in climate, such as the reduced reflectivity of the Earth's surface when ice melts)

ppm: parts per million

ppt: parts per thousand

practical salinity unit (PSU): a unitless measure of salinity equal to parts per thousand

precession: in the context of Milankovitch Cycles, the variation in the direction at which the Earth's rotational axis is pointing

pressure ridge: jagged ridges created from colliding and buckling ice floes (14.1)

primary production: the synthesis of organic compounds from aqueous carbon dioxide by plants, algae, and bacteria

protoplanetary disk: a rotating cloud of gas and dust surrounding a young star

pycnocline: a region in the water column where there is a large change in density over a small change in depth

quartz: a mineral composed of silicon and oxygen atoms in the ratio of 1 Si:2 O; one of the most abundant minerals in the Earth's surface

radiation: the emission of energy in the form of electromagnetic waves

radiolarian: microscopic (0.1 to 0.2 mm) marine protozoa that produce silica shells

rain shadow: arid conditions behind a mountain range, as rising air on the other side of the mountain caused rain, leaving only dry air to descend back down the mountain

reef: a mound of carbonate formed in shallow tropical marine environments by corals, algae and a wide range of other organisms

regenerated production: primary production resulting from the recycling of nutrients within an ecosystem

remnant magnetism: magnetism of a body of rock that formed at the time the rock formed and is consistent with the magnetic field orientation that existed at that time and place

residence time: the average amount of time an element will remain in the ocean before being removed

restoring force: the force that opposes a wave-generating force and attempts to return the sea surface to the still water level

ridge push: the concept that at least part of the mechanism of plate motion is the push of oceanic lithosphere down from a ridge area

rift valley: a valley created when crust subsides along a divergent plate boundary

rip current: a strong flow of water outward from a beach

rogue wave: an exceptionally large wave arising among a series of smaller waves

rule of constant proportions: the major ions in seawater are always found in the same proportions, regardless of overall salinity

runoff: flow of water down a slope, either across the ground surface, or within a series of channels

s-wave: a seismic body wave that is characterized by deformation of the rock perpendicular to the direction that the wave is propagating

salinity: the concentration of dissolved ions in water

salt wedge estuary: an estuary with mostly fresh surface water, and a wedge of seawater intruding along the bottom

sand: a mineral or rock fragment ranging in size from 1/16th to 2 mm

saturation: the amount of a substance currently dissolved in the water, relative to the maximum possible content

scarp: a short, steep wall carved out by wave action between the foreshore and the berm of a beach

scavenged element: elements whose vertical profiles show high abundance at the surface and declining concentrations at depth as they are removed by sinking particles

sea breeze: winds blowing from the ocean towards the land

sea cave: a shallow cave formed on a rocky shore by wave erosion

sea cliff: a coastal escarpment that is typically eroding inland as a result of wave action

sea ice: ice formed from the freezing of seawater

sea stack: a prominent rocky island that is a remnant of the erosion of a headland

sea state: describes the current wave conditions in an area

seafloor spreading: the formation of new oceanic crust by volcanism at a divergent plate boundary

seamount: a submerged mountain rising from the seafloor

seawall: a wall built against a sea cliff or dune to prevent erosion from wave action

sediments: unconsolidated particles of mineral or rock that settle to the seafloor

seismic: pertaining to earthquakes

seismology: the study of vibrations within the Earth

semidiurnal tide: a tidal cycle with two high and two low tides per day, each of roughly equal heights

shallow water wave: a wave in water with a depth less than 1/20 of the wavelength

shelf break: the boundary between the continental shelf and continental slope, where the angle of the seafloor begins to get steeper

significant wave height (Hs): the mean height of the largest one-third of the waves in a wave spectrum

silica: a form of the mineral quartz

siliceous sediment: sediment dominated by particles of silica, often from the shells of marine organisms

silt: sedimentary particles ranging is size from 1/256th to 1/16th of a mm

slab pull: the concept that at least part of the mechanism of plate motion is the pull of oceanic lithosphere down into the mantle

slack tide: period of little water movement between an incoming and outgoing tide

slightly stratified estuary: where salinity increases from the head to the mouth, but there is also a slight increase in salinity with depth at any point; also called a partially mixed estuary

snow line: in astronomy the radius around a star at which represents the boundary between gases (or liquids) and solids

SOFAR channel: range of depths around 1000 m where sound travels the slowest, so sound waves are refracted back into the channel and can be propagated long distances

solar system: a star and the planets surrounding it

solar wind: a stream of ionized (charged) particles away from the Sun

solubility: the amount of a dissolved substance that water can hold under a particular set of conditions, which are usually defined as 0 degrees C and 1 atmosphere of pressure

sonar: acronym for sound navigation and ranging; a method of using sound echoes to detect objects

sorting: how uniform the particles of a sediment are in terms of size

sounding: a single measurement of ocean depth

specific heat: the heat required to raise the temperature of 1 g of a substance by 1 degree C

spherule: a microscopic piece of space dust

spilling breaker: a breaker on relatively flat beaches that slowly increases its height and collapses

spit: a sand or coarser deposit extending from shore out into open water

splash wave: a wave formed when something falls into the ocean and creates a splash

spring tide: the period of maximum tidal range when the moon, sun and Earth are aligned

stable gas: unreactive gases that dissolve in seawater

steady state: where a system shows no net change, as input equals output

still water level: where the water surface would be if there were no waves present and the sea was completely calm

storm surge: an area of high water that moves with storm systems

subducted: when part of a plate is forced beneath another plate along a subduction zone

subduction zone: the sloping region along which a tectonic plate descends into the mantle beneath another plate

subittoral zone: the region of a coast from the low tide line to the end of the continental shelf

supralittoral zone: the region of a coast above the high tide line

surf beat: an irregular surface wave pattern caused by mixed interference

surf zone: the near-shore zone where waves are breaking into surf

surface tension: where a cohesive layer forms on the water surface due to attraction between water molecules

surging breaker: waves that break on steep beaches with a very sudden increase in height and sudden collapse right on the beach

swash: the upward motion of a wave on a beach (typically takes place at the same angle that the waves are approaching the shore)

swell: regular, long-period waves that have sorted themselves based on speed

tablemount: a flat-topped seamount (also called a guyot)

tabular iceberg: flat-topped, steep-sided iceberg with a length greater than five times the height

tectonic estuary: an estuary formed from flooding following the tectonic subsidence of land

tectonic plate: a region of the lithosphere that is considered to be moving across the surface of the Earth as a single unit

tektite: solidified glass fragments ejected during meteorite impacts

terrestrial planet: a planet with a rocky mantle and crust and metallic core (e.g., Earth)

terrigenous sediment: referring to sedimentary particles that originated on a continent

test: the shell-like hard parts (either silica or carbonate) of small organisms such as radiolarians and foraminifera

thermal expansion: the increase in the volume of a body water as its temperature rises and its density decreases

thermocline: a region in the water column where there is a dramatic change in temperature over a small change in depth

thermohaline circulation: deep ocean circulation driven by differences in water density

tidal bore: a wave that moves up a river with an incoming tide

tidal day: the amount of time between a tide on one day and the same tide the following day

tidal range: the difference in height between the high and low tides

tidal volume / tidal prism: the volume difference of an area between low and high tides

tombolo: a sand or coarser deposit connecting an island or rocky prominence to a larger body of land

trade winds: prevailing wind bands between the equator and 30 degrees latitude

transform boundary: a boundary between two plates that are moving horizontally with respect to each other

transform fault: a type of fault in which two pieces of crust slide past one another

trough: the lowest point of a wave

tsunami: a long-wavelength wave produced by the vertical motion of the floor of the ocean, typically related either to an earthquake or other submarine seismic event

turbidity current: a current moving down downhill along the bottom, driven by the weight of the sediment within it

twilight zone: depths of the water column where there is some light penetration, but not enough to support photosynthesis; corresponds to the mesopelagic zone, 200-1000 m. Also known as the dysphotic zone

upwelling: process by which deeper water is brought to the surface

vertically-mixed estuary: estuary with complete mixing of fresh and salt water, where salinity is constant at all depths in a particular location but increases towards the estuary mouth; also called a well-mixed estuary

water mass: a volume of seawater with a distinctive density as a result of its unique profile of temperature and salinity

wave base: the depth of water that is affected by the sub-surface orbital motion of wave action (approximately one-half of the wavelength)

wave height: the distance between the crest and trough of a wave

wave steepness: the ratio of wave height to wavelength

wave-cut platform/terrace: a nearly-horizontal bench of rock eroded by waves within the surf zone

wavelength: the distance between the crests of two waves

weathering: a range of processes taking place in the surface environment, through which solid rock is transformed into sediment and ions in solution

wedge iceberg: iceberg with a steep face next to a more gradually sloping side

well-mixed estuary: estuary with complete mixing of fresh and salt water, where salinity is constant at all depths in a particular location but increases towards the estuary mouth; also called a vertically-mixed estuary

westerlies: the dominant wind bands between 30 and 60 degrees latitude in each hemisphere

western intensification: currents on the western side of a gyre are faster, deeper, and narrower than currents on the eastern side

zooplankton: small, drifting carnivorous organisms

INDEX